Pioneering the Cosmos

Second Edition of:

From Octonion Cosmology to the Unified SuperStandard Theory of Particles

Stephen Blaha Ph. D.

Blaha Research

Detailed Analysis of 10 Octonion Spaces, Superverse & Their Particles
Octonion Spaces Computation Model
Strings, Bags, Data Spaces, Minispaces, Higgs Bosons
QUeST Direct Connection to the Unified SuperStandard Theory (UST)
Parallels to *Sefer Yetzirah* Cosmology, Linguistic Theory & God-View
Sefer Yetzirah Hints of Lost Ancient Physics
The 231 Gates to Space-Time
A God-View Based on Logic Using Monads
Cascading Quantum Probability Tweaks to Manage History
Life After Death Through Monads?

Pingree-Hill Publishing

MMXXI

ISBN: 978-1-7356795-5-6

Rev. 00/00/01 February 23, 2021

To Margaret

Some Other Books by Stephen Blaha

All the Megaverse! Starships Exploring the Endless Universes of the Cosmos using the Baryonic Force (Blaha Research, Auburn, NH, 2014)

SuperCivilizations: Civilizations as Superorganisms (McMann-Fisher Publishing, Auburn, NH, 2010)

All the Universe! Faster Than Light Tachyon Quark Starships & Particle Accelerators with the LHC as a Prototype Starship Drive Scientific Edition (Pingree-Hill Publishing, Auburn, NH, 2011).

Unification of God Theory and Unified SuperStandard Model THIRD EDITION (Pingree Hill Publishing, Auburn, NH, 2018).

The Exact QED Calculation of the Fine Structure Constant Implies ALL 4D Universes have the Same Physics/Life Prospects (Pingree Hill Publishing, Auburn, NH, 2019).

Unified SuperStandard Theory and the SuperUniverse Model: The Foundation of Science (Pingree Hill Publishing, Auburn, NH, 2018).

Quaternion Unified SuperStandard Theory (The QUeST) and Megaverse Octonion SuperStandard Theory (MOST) (Pingree Hill Publishing, Auburn, NH, 2020).

Unified SuperStandard Theories for Quaternion Universes & The Octonion Megaverse (Pingree Hill Publishing, Auburn, NH, 2020).

The Essence of Eternity: Quaternion & Octonion SuperStandard Theories (Pingree Hill Publishing, Auburn, NH, 2020).

A Very Conscious Universe (Pingree Hill Publishing, Auburn, NH, 2020).

From Octonion Cosmology to the Unified SuperStandard Theory of Particles (Pingree Hill Publishing, Auburn, NH, 2020).

Available on Amazon.com, bn.com Amazon.co.uk and other international web sites as well as at better bookstores (through Ingram Distributors).

CONTENTS

INTRODUCTION 1

0. HYPERCOMPLEX SPACES - EXTENSION OF THE UNIFIED SUPERSTANDARD THEORY (UST) TO HIGHER DIMENSIONS 3

- 0.1 SUGGESTIONS OF A DEEPER BASIS FOR UST 3
- 0.2 HYPERCOMPLEX NUMBERS 3
- 0.3 MATHEMATICAL PICTURE LANGUAGE 4
- 0.4 NUMBER DESCRIBES EVERYTHING 4
- 0.5 AXIOMS FOR THE UNIFIED SUPERSTANDARD THEORIES 5
- 0.6 GENERAL IMPLICATIONS OF THE AXIOMS 6
- 0.7 HYPERCOMPLEX NUMBER BASED HIGHER DIMENSIONS 8
- 0.8 FROM COMPLEX COORDINATES TO HYPERCOMPLEX COORDINATES 9
- 0.9 QUATERNION COORDINATE SYSTEMS 9
- 0.10 OCTONION COORDINATE SYSTEMS 10
- 0.11 "COMPOSITE" OCTONION COORDINATE SYSTEMS 11

APPENDIX 0-A. QUATERNION FEATURES 13

- 0-A.1 SOME BASIC QUATERNION FEATURES 13

APPENDIX 0-B. SOME OCTONION FEATURES 15

1. THE SPECTRUM OF 10 OCTONION SPACES 17

- 1.1 FUNDAMENTAL AXIOMS 17
- 1.2 INSTANCES OF THE OCTONION SPACES 19
- 1.3 NESTING OF OCTONION SPACES 19
- 1.4 MATCH OF LEVEL 6 SPACE WITH UST AND LEVEL 5 WITH UTMOST 20
- 1.5 OCTET SPACE 7 20
- 1.6 THE SPECTRUM OF PHYSICAL OCTONION SPACES 20
- 1.3 CHARACTERISTICS AND ROLES OF THE SPACES 22
  - *1.3.0 Superverse Space* 23
  - *1.3.1 Spaceless Spaces (1 – 3)* 23
  - *1.3.2 Octonion Octonion Maxiverse Space (4)* 23
  - *1.3.3 Quaternion Octonion Megaverse Space (5)* 23
  - *1.3.4 Complex Octonion Universe Space (6)* 23
  - *1.3.5 Quaternion Point Universe Minispace (7)* 24
  - *1.3.6 Quaternion Point Universe Minispace (8 and 9)* 24
  - *1.3.7 Quaternion Point Universe Minispace (10)* 24
  - *1.3.8 Comments* 24
- 1.4 THE PATTERN OF THE TEN SPACES 25
  - *1.4.1 Unification of the 10 Spaces in the Superverse* 25
- 1.5 COSMOLOGICAL EVOLUTION 27

2. QUEST FOR UNIVERSES 31

- 2.1 COMPLEX OCTONION SPACE OF COMPLEX OCTONION DIMENSIONS 31
  - *2.1.1 Layers* 33
  - *2.1.2 Fundamental Group Representations* 33
  - *2.1.3 Partition of 32 × 8 Dimension Array* 33

2.1.4 Map Between Octonion Cosmology and UST *36*
2.3 QUEST SYMMETRIES 36
2.3.1 QUeST Internal Symmetry Groups *36*
2..3.2 QUeST Space-Time *36*
2.3.3 QUeST/UST Fermion Spectrum *36*
2.3.4 Generation and Layer groups of UST, QUeST and BQUeST *36*
2.3.5 Fermion- Dimension Duality *36*
2.4 16 DIMENSION BLOCK PATTERN OF FERMIONS 39

3. THE DERIVATION OF QUATERNION QUEST UNIVERSES 41

3.1 BQUEST ORIGIN 41
3.2 A MODEL DYNAMICS OF THE SEED – ANTI-SEED GENERATION OF A QUEST UNIVERSE 42
3.2.1 U(16) Species Group Symmetry for Dirac Matrix Indices *42*
3.3 THE INTERACTION GENERATING A UNIVERSE PARTICLE 44
3.4 U(16) SYMMETRY OF A_I 45
3.5 INDICES OF $[A_{1AB}]$ TRANSFORMED TO INTERNAL DIMENSIONS 46
3.6 THE DIFFERENCE BETWEEN INDICES AND INTERNAL DIMENSIONS 46
3.7 ORIGIN OF UNIVERSE 46
3.8 NATURE OF THE SEED FERMION AND ANTIFERMION 47
3.9 DEMONSTRATION THAT UTMOST FERMIONS OCCUR IN FOUR SPECIES 47

4. CONNECTION OF SEED FERMION STRUCTURE TO QUEST DIMENSIONS STRUCTURE 49

4.1 STRUCTURE OF 16-SPINOR U AND V 49
4.2 COMPRESSION OF U AND V INTO COMPOSITE SPINORS 49
4.3 BLOCK FORM OF $[U_{BA}]$ ARRAY 51
4.4 QUEST DIMENSION ARRAY INPLIED BY 32×8 SPINOR ARRAY FORM 53
4.5 PROGRESSION FROM BQUEST TO UST 56

5. QUEST UNIVERSE STRUCTURE & ITS RELATION TO UST 57

5.1 INTERNAL SYMMETRIES 57
5.2 QUEST → UST SYMMETRIES 60
5.3 U(4) GENERATION, U(4) LAYER AND U(1) FERMION GROUPS 61
5.4 THE GENERATION GROUP 61
5.5 THE LAYER GROUP 62
5.6 THE FERMION GROUPS 63
5.7 PARTICLE SPECTRUMS OF QUEST: FERMIONS, AND VECTOR BOSONS 63
5.7.1 Vector Bosons *64*
5.7.2 Fermions *64*
5.8 OTHER QUEST-UST TOPICS 64

6. UTMOST FOR THE MEGAVERSE 67

6.1 UTMOST SPACE AND INTERNAL SYMMETRIES 67
6.2 UTMOST SYMMETRIES 72
6.2.1 UTMOST Internal Symmetry Groups *72*
6.2.2 UTMOST Space-Time *72*
6.2.3 UTMOST Fermion Spectrum *72*
6.3 UTMOST FERMIONS 72

7. THE BMOST DERIVATION OF THE UTMOST MEGAVERSE 75

7.1 BQUEST ORIGIN 75
7.2 A MODEL DYNAMICS OF THE SEED – ANTI-SEED GENERATION OF THE MEGAVERSE 76

7.2.1 U(32) Species Group Symmetry for Dirac Matrix Indices 77
7.3 THE INTERACTION GENERATING A MEGAVERSE PARTICLE 78
7.4 U(16) SYMMETRY OF A_I 79
7.5 INDICES OF $[A_{1AB}]$ TRANSFORMED TO INTERNAL DIMENSIONS 80
7.6 THE DIFFERENCE BETWEEN INDICES AND INTERNAL DIMENSIONS 80
7.7 ORIGIN OF THE MEGAVERSE 80
7.8 NATURE OF THE SEED URFERMION AND ANTIURFERMION 81

8. CONNECTION OF SEED URFERMION STRUCTURE TO UTMOST DIMENSIONS STRUCTURE 83

8.1 STRUCTURE OF 32-SPINOR U AND V 83
8.2 COMPRESSION OF U AND V INTO COMPOSITE SPINORS 83
8.3 BLOCK FORM OF $[U_{BA}]$ ARRAY 86

9. THE BMOST – UTMOST CONNECTION 89

10. MAXIVERSE OCTONION OCTONION SPACE: 10 SPACE-TIME DIMENSIONS – WITH 32-SPINOR URFERMIONS 91

10.1 BLOCK STRUCTURE OF UNIVERSE, MEGAVERSE AND MAXIVERSE SPACES 93

11. THE THREE SPACELESS SPACES 95

11.1 COMPUTATION AND A THREE SPACELESS SPACES PARADIGM 95
11.2 ANALOGUE TO THE *SEFER YETZIRAH* AND THE LOGICAL GOD-VIEW IN CHAPTER 17 97

12. THE MINISPACES 99

13. THE SUPERVERSE SPACE 101

14. STRINGS IN THE MAXIVERSE 103

15. BAGS AND MINIVERSES 105

16. HIGGS BOSON SECTOR OF OCTONION COSMOLOGY 107

16.1 THE NUMBER OF HIGGS BOSONS IN QUEST AND UTMOST 107
16.2 HIGGS AND SYMMETRY BREAKING 107

17. OCTONION COSMOLOGY PARALLELS IN *SEFER YETZIRAH* COSMOLOGY, LINGUISTICS, AND GOD-VIEW 109

17.1 AN EXEGESIS OF THE *SEFER YETZIRAH* 109
17.2 PHYSICAL CONCEPTS IN THE *SEFER YETZIRAH* 109
17.3 COSMOLOGY OF *SEFER YETZIRAH* 110
17.4 COMPARISON OF ITS COSMOLOGY WITH OCTONION COSMOLOGY 111
17.5 SYNCRETIC FORMULATION OF *SEFER YETZIRAH* AND OCTONION COSMOLOGY 111
17.5.1 Sefirot Groupings *113*
17.6 ORIGIN OF *SEFER YETZIRAH* COSMOLOGY 115
17.7 PARALLELS IN LINGUISTICS WITH OCTONION COSMOLOGY 115
17.7.1 The 231 Gates in the Wall *116*
17.8 PARALLEL OF *SEFER YETZIROT,* AND LATER SCHOOLS, GOD-VIEW TO OCTONION COSMOLOGY.. 117
17.9 PARALLELS TO *SEFER YETZIRAH* COSMOLOGY, LINGUISTIC THEORY & GOD-VIEW 117
17.10 SEFER YETZIRAH HINTS OF LOST ANCIENT PHYSICS 117
17.10.1 The 231 Gates to Space-Time *117*
17.11 CASCADING QUANTUM PROBABILITY TWEAKS TO MANAGE HISTORY ASSUMING GOD 118

17.11.1 Quantum Source of Free Will *118*
17.12 A Logical View of God 118

18. MONADS OF ELEMENTARY PARTICLES – EXTENSION TO 10 SPACES 129

18.1 Internal Symmetry Functionals Space F_{IS} 130
18.2 Spin-Momentum Functionals Space F_{SP} 130
18.3 Fourier Wave Space F_{FW} 130
18.4 Space-Time and Functional Spaces 130
18.5 Ten Spaces of Octonion Cosmology 130
18.6 Dynamic Generation of Quantum State Functionals 130
18.7 Monads and The Possibility of a Form of Life After Death 131

APPENDIX 18-A. QUANTUM ENTANGLEMENT AND ACTION-AT-A-DISTANCE 133

18-A.1 The EPR Two System State Example 133

APPENDIX 18-B. WHY FACTORIZE QUANTUM FIELDS INTO A FUNCTIONAL AND A FOURIER EXPANSION? 135

19. OCTONION COSMOLOGY 137

APPENDIX A. ENHANCED QUANTUM FIELD THEORY: TWO TIER & PSEUDOQUANTUM FROM CHAPTERS 40 AND 41 OF BLAHA (2020C) 139

40.1 Two-Tier Features in 4-Dimensional Space-Time 140
40.2 Simple Two-Tier X^{μ} Formalism 142
40.3 Y^{μ} Gauge 143
40.4 Scalar Field Quantization Using X^{μ} 145
40.5 Scalar Feynman Propagators 146
40.6 String-like Substructure of the Theory 147
40.7 Two-Tier Complexon Quantum Fields 148
40.8 Complexon Feynman Propagator 150
40.9 Vacuum Fluctuations 150
40.10 Time Intervals in General Relativity 151
40.11 Vacuum Fluctuations in the Gravitation Fields 152
40.12 Two-Tier Features in D-Dimensional Space-Time (such as the Megaverse) 153
41.1 General Case of PseudoQuantization in Differing Coordinate Systems 157
41.2 Two-Tier PseudoQuantum Field Theory 158
41.3 PseudoQuantum Higgs Scalar Particle Field Theory in D-dimensional Space-Time .. 159
41.3.1 The Enigma of Higgs Particles and the Higgs Mechanism *159*
41.3.2 PseudoQuantization of Scalar Particles *161*
41.3.3 Vacuum States for Scalar (Higgs) Particles with Non-Zero Vacuum Expectation Values ... *162*
41.3.4 Interpretation of Negative Energy Scalar Particle States *164*
41.3.5 Contrast with Conventional Second Quantization of Scalar Particles *164*
41.3.6 Why Inertial Reference Frames are Special *165*
41.3.7 PseudoQuantization Reveals More Physical Consequences than the Higgs Mechanism of Scalar Particles *165*
41.3.8 The T Invariance Issues of Our PseudoQuantized Scalar Particle Theory *166*
41.3.9 Retarded Propagators for Our Quantized Higgs Particles *167*
41.3.10 The Local Arrow of Time *168*
41.3.11 Space-Time Dependent Particle Masses *168*
41.3.12 Inertial Mass Equals Gravitational Mass *169*
41.3.13 Benefits of the PseudoQuantization Method *170*

41.4 Two-Tier Formulation and PseudoQuantization in the Megaverse 171

APPENDIX B. COMPLEX GENERAL RELATIVITY & DIMENSION – INDEX CONNECTION FROM CHAPTERS 50 AND 51 OF BLAHA (2020C) 173

50.1 Tetrad (Vierbein) Formalism 173
50.2 Complex General Relativistic Transformations 174
50.3 Structure of Complex General Coordinate Transformations 175
50.4 Complex Affine Connection – General Relativistic Reality Group 176
51.1 Species Group Covariance 179
51.2 Spontaneous Symmetry Breaking of the General Relativity U(4) Reality Group – The Species Group 180
51.3 Species Group Higgs Mechanism Contributions to Fermion Masses 181
51.4 Species Group Higgs Masses Shows Inertial Mass Equals Gravitational Mass 182

APPENDIX C. SOME FEATURES OF QUEST QUATERNION SPACE 183

C.1 Thirty-Two Complex Quaternion Space – 32 × 8 Dimension Array 183
C.1.1 QUeST Internal Symmetry Groups *184*
C..1.2 QUeST Space-Time *184*
C.1.3 Generation and Layer groups of UST, QUeST and BQUeST *184*
C.1.4 Fermion- Dimension Duality *184*
C.1.5 Fermion Structure Extracted from QUeST Symmetry Structure *184*
C.2 Thirty-Two Complex Quaternion Space – 16 × 16 Dimension Array 189
C.3 A Partition to Real 3+1 Dimension Space-times 195

APPENDIX D. EVIDENCE FOR UNIVERSE PARTICLES 197

D.1 Evidence for Entities Beyond the Universe 197
D.1.1 Great Attractors *197*
D.1.2 Bright Bumps in Universe Suggesting Collision with Another Universe *197*
D.1.3 Cold Spot in Universe Suggesting Collision with Another Universe *198*
D.1.4 Conclusion *198*

APPENDIX E. SOME FEATURES OF UTMOST 199

E.1 UTMOST Space 199
E.2 UTMOST Fermions 204
E.3 Partition of UTMOST into QUeST Subspaces 206

APPENDIX F. "D" DIMENSION QUANTUM FIELD THEORY FROM CHAPTER 10 OF BLAHA (2017C) 209

10.2 The Particle Interpretation of Extended Wheeler-DeWitt Equation Solutions 209
10.3 "Free Field" Dynamics of Fermionic Universe Particles 210
10.3.1 Four Types of Fermionic Universe Particles *211*
10.3.1.1 Dirac-like Equation – Type I universe Particle 212
10.3.1.2 Complex Boosts 213
10.3.1.3 Tachyon Universe particle Dirac Equation 214
10.3.1.4 Type IIa Case: Left-Handed Tachyonic Universe Particles 214
10.3.1.5 Type IIb Case: Right-Handed Tachyonic Universe Particles 215
10.3.1.6 Type III Case: "Up-Quark-like" Universe Particles 216
10.3.1.7 Type IVa Case: Left-Handed "Down-Quark-like" Tachyonic Universe Particles 217
10.3.1.8 Type IVb Case: Right-Handed Down-Quark-like Tachyonic Universe Particles 218
10.3.2 Lagrangians *219*
10.3.2.1 Type I Universe Particle Lagrangian 219
10.3.2.2 Type II Tachyon Universe Particle Lagrangian 219
10.3.2.3 Type III "Up-Quark-like" Universe Particle Lagrangian 219

10.3.2.4 Type IV "Down-Quark-like" Tachyon Universe Particle Lagrangian 220
10.3.3 Form of The Megaverse Quantum Coordinates Gauge Field *220*
10.3.3.1 Y^μ Fock Space Imaginary Coordinate States 222
10.3.3.2 Y_u Coherent Imaginary Coordinate States 223
10.3.3.3 Quantization of the Type I Free Universe Particle Dirac Field 224
10.3.3.4 Feynman Propagators for the Type I Free Universe Particle Dirac Field....225
10.3.3.5 Feynman Propagators for the Types II, III, and IV Free Universe Particle Dirac Fields....226
10.3.4 Expanding and Contracting Universes: Impact of Time Dependent Universe Particle Masses *226*

REFERENCES 229

INDEX 237

ABOUT THE AUTHOR 241

FIGURES and TABLES

Figure 0.1. The Pythagorean tetrackys, a psiphi representation of 10............................ 4
Figure 0.2. Psiphi diagram of the dimensions of 3 + 1 complex quaternion space. Each row represents a complex quaternion. Note the coordinates of eq. 0.4 map directly to dimensions represented by •'s. 10
Figure 1.1. The levels and dimensions of the octonion space octet. Note the lowest level is octonionic as it should be.. 19
Figure 1.2. Nesting of octonion spaces.. 19
Figure 1.3. The ten spaces with their spectrum number and their space-time dimensions indicated within each oval as in Fig. 1.4. The number of spinor components for each fermion - antifermion pair that annihilates to produce the "next space down" is specified next to each arrow for the lower spaces. On the right are symbols for instances (particles) of spaces. 21
Figure 1.4. The spectrum of the Superverse and the ten octonion spaces. The spaces are numbered from 0 through 10. The numbers in parentheses in column 2 are the number of dimensions in each coordinate. The items in column 3 are the number of rows of dimensions (1024, 512, 256, 128, 64, 32, 16, 4, 4, 4,4). 22
Figure 1.5. The descent of the ten spaces with their spectrum number and their space-time dimensions indicated within each oval. The number of spinor components for each fermion - antifermion pair that annihilates to produce the "next space down" is specified next to each arrow for the seven lower spaces. 26
Figure 1.6. The 10 spaces of Fig. 1.4 in block form within a 1024 by 1024 dimension space – the Superverse space. The 10 spaces are 512 by 512, 256 by 256, …, 4 by 4. The shapes of the spaces diagrammed in this figure is not significant.......................... 27
Figure 1.7. The five spaces relevant for the physical Cosmos. The three colored regions have non-zero dimension, and have a non-trivial space containing fermions and bosons with interactions.. 29
Figure 2.1. The 16 × 16 dimension array used by QUeST. Each • represents a dimension.. 32
Figure 2.2. The 32 octonions form of the dimension array subdivided into 4 layers of 8 rows..Each layer will be seen to map to a block of fundamental group representations as shown in Figs. 2.3 and 2.4. 32
Figure 2.3.. Map between fundamental group representations and their dimensions. (Dimensions can also be represented as complex.) 33
Figure 2.4. The four layers of QUeST internal symmetry groups (and space-time) for the 32 octonion dimension form of space. Note: each row has an 8 • octonion. Note the left column of blocks combine to specify a 4 dimension octonion space-time. Note each

layer requires 64 dimensions.. *Note the duplication (using the "Two" label) of each symmetry in each layer.* 34
Figure 2.5. Sets of 16 dimension blocks for two of four layers of the 32 × 8 array. Each "dashed" block is a 4 × 4 = 16 array of dimensions. The solid line separates the 64 dimension blocks. This set of 8 blocks contains the 8×16 = 128 dimensions of two of the four layers. 35
Figure 2.6. Fermion particle spectrum and partial examples of the pattern of mass mixing of the Generation group and of the Layer group. Unshaded parts are the known fermions with an additional, as yet not found, 4th generation. The lines on the left side (only shown for one layer) display the Generation mixing within each layer. The Generation mixing occurs within each layer using a separate Generation group for each layer. The lines on the right side show Layer group mixing (for Dark matter) with the mixing among all four layers for each of the four generations individually. There are four Layer groups for Normal matter and four Layer groups for Dark matter.. There are 256 fundamental fermions. QUeST and UST have the same fermion spectrum.......... 37
Figure 2.7. Fundamental fermions have a 1:1 correspondence with QUeST dimensions. Note the number of dimensions in each row is 8 – the number of dimensions in an octonion. Correspondingly the number of fermions in each row is 8 – a suggestive similarity. Each layer has four normal fermion generations and four Dark fermion generations. Each dot (pebble) represents a dimension in the left part of the figure and a fermion in the right part. 38
Figure 2.8. A block form of a 16 × 16 QUeST fermion array with each block row corresponding to one layer. Each block contains four generations of fermions. The result is 4 × 4 blocks. The label e q-up indicates a charged lepton – up-type quark pair, ν q-down indicates a neutral lepton – down-type quark pair, and so on. Note the blocks can be reaaranged into a 32 × 8 form without physical consequences at this level of discussion since the right two columns and the lowest two rows are all Dark at present. 39
Figure 3.1. Diagram for the transition of seed - antiseed fermions to a universe particle. 44
Figure 4.1. The 16 spinors of a 8 dimension spinor. Each spinor has 16 rows. 49
Figure 4.2. Composite u-type and v-type spinors illustrating the structure of each type of spinor based on combining the two columns of each spinor type into one column. Each spinor is subdivided into 4 row sections. The third and fifth columns are symbols for each sector. 50
Figure 4.3. Outer product array $[U_{ba}]$ (eq. 1.9) of the composite u-type and v-type spinors illustrating the structure of the outer product array of uv's. 50
Figure 4.4. Block form of the 16 × 16 $[U_{ba}]$ array. This is also the form of the QUeST dimension array of 256 dimensions. The blocks are divided by dashed lines that

separate 64 dimension sections. These sections map to layers in QUeST (UST) as shown in Figs. C.2 and C.3. 51
Figure 4.5. Fig. A: Block form of the 16 × 16 [U_{ba}] array arranged to a 32 × 8 form to display a layer structure similar to QUeST. Fig. B: The layers were grouped by sector type and ordered from "large" spinor components to "small" spinor components. 52
Figure 4.6. The 32 octonion dimensions QUeST array subdivided into 4 layers of 8 rows..Each layer will be seen to map to a block of fundamental group representations as shown in Figs. 4.7 and 4.8. 53
Figure 4.7. The four layers of QUeST internal symmetry groups (and space-time) for 32 dimension complex quaternion space. Note: each row has an 8 pebble • octonion. Note the left column of blocks combine to specify a 4 dimension octonion space-time. Note each layer has 64 dimensions. 54
Figure 4.8. Sets of 16 dimension blocks for the first two layers (of four) of the 32 × 8 array. Each "dashed" block is a 4 × 4 = 16 array of dimensions. This set of 8 blocks contains the 8×16 = 128 dimensions of layer 1 and Layer 2. 55
Figure 4.9. The seed fermion spinor array structure is mirrored by the QUeST dimension structure. This is evidence for the seed origin of QUeST via BQUeST. 56
Figure 5.1. Pattern of symmetry breaking of QUeST U(128). 57
Figure 5.2. The four layers of QUeST internal symmetry groups (and space-time) for its 32 dimension octonion space. Note: each row has an 8 • octonion. Note the left column of blocks combine to specify a 4 dimension octonion space-time. Note each layer requires 64 dimensions. 59
Figure 5.3. Four layers of Internal Symmetry groups in QUeST. The groups in each layer are independent of those in other layers. The groups in each block of each layer are independent of those in the other blocks. Each block contains 16 dimensions. The block dimensions furnish fundamental representations for the groups listed. The entire set of blocks contains 256 dimensions. Each layer contains 64 dimensions. The first two columns are for the "Normal" sector. The last two columns are for the "Dark" sector (although most of the Normal sector is Dark observationally at present.) This Figure also holds for UST with the addition of Fermion groups. 60
Figure 6.1. The 32 quaternion octonion dimension UTMOST array. It is a 32 × 32 dimension array of •'s. It has 1024 dimensions. It is grouped into four layers in rows of eight. 67
Table 6.2. Map between fundamental representations and their dimensions. 68
Figure 6.3. The internal symmetry groups of *one layer* (consisting of 8 rows in Fig. 6.1) of the four layers of 32 × 32 dimension UTMOST. The other three layers are copies of the this layer. Note the Fermion $U(1)^4$ groups. 69
Figure 6.4 The *first* of the four layers of UTMOST dimensions with boxes around sets of dimensions for fundamental group representations 70

Figure 6.5. Four layers (each in two rows) in the 32 × 32 dimension UTMOST array composed of 4 × 4 blocks, which are within the four block 8 × 8 sections for each pair: Normal+Dark1, Dark2+Dark3, Dark4+Dark5 and Dark6+Dark7. In total they form the 32 × 32 = 1024 UTMOST dimension array.......... 71
Figure 6.6. Spectrum of UTMOST fermions in a 16×64 format. Each fermion is represented by a •..Each set of eight •.'s represents a charged lepton, a neutral lepton, three up-type quarks, and three down-type quarks. There are eight sets of four species in four generations which are in turn in 4 layers. There are 1024 fundamental fermions taking account of quark triplets. 73
Figure 6.7. Block form of the 32 × 32 UTMOST fermion array with each row corresponding to *half of an UTMOST layer*. Thus 8 × ½ = 4 layers results. Each block contains four generations of fermions. The result is sixty-four 4 × 4 blocks. The label e q-up indicates a charged lepton – up-type quark pair, ν q-down indicates a neutral lepton – down-type quark pair, and so on. *The form displayed here explains why generations come in fours.* 74
Figure 7.1. Diagram for the transition of seed - antiseed urfermions to a Megaverse particle. 78
Figure 8.1. The 32 spinors of a 10 dimension spinor. Each spinor has 32 rows. 83
Figure 8.2. Composite u-type and v-type spinors illustrating the structure of each type of spinor. Each spinor is subdivided into 4 sections of 8 components. The third and fifth columns are symbols for each sector. 84
Figure 8.3. Outer 32 × 32 product array [U_{ba}] (eq. 7.9) of the composite u-type and v-type spinors illustrating the structure of the outer product array of uv's. Each column represents 8 spinor columns, making 32 columns in all. 85
Figure 8.4. Block form of the 32 × 32 [U_{ba}] array. This is also the form of the UTMOST dimension array of 1024 dimensions. The sets of four blocks are divided by dashed lines that specify 256 dimension sections. These sections map to layers in UTMOST as shown in Figs. E.3, E.4 and E.5 for one of the four UTMOST layers. Each UTMOST layer is equivalent to a QUeST 256 dimension array.......... 86
Figure 8.5 The 256 dimension groups of the four 8 × 8 blocks in each section of Fig. 8.4 Each 256 dimension section is one of the four layers of UTMOST dimensions. 87
Figure 10.1. One layer of the four layers of one of the four copies of UTMOST dimensions *plus* an additional 12 dimension block with boxes around sets of dimensions for fundamental group representations. There are 16 copies of the internal symmetries listed here in Octonion Octonion space. Compare this figure to the UTMOST Fig. E.4 to see details of Normal and Dark sectors indicated above. 92
Figure 10.2. A 64 dimension (64-block) block. It is one layer in QUeST. 93
Figure 10.3. A table of the 16-blocks and 64-blocks in QUeST, UTMOST, and Maxiverse space. The additions for the Maxiverse result from using a 10 quaternion

octonion space-time as shown sbove. The quantities above increase by factors of four except Maxiverse space where the quantities have additions due to the choice of space-time dimension.......... 94
Figure 10.4. A table of the 16-blocks and 64-blocks of the three space-time-less spaces. 94
Figure 10.5. A table of the 16-blocks and 64-blocks of the Superverse space.......... 94
Figure 17.1. Map between *Sefer Yetzirah* and octonion spaces of Fig. 1.4.......... 111
Figure 17.2. The Superverse and the spectrum of ten octonion spaces shown in Fig. 1.4. In total they are 11 spaces and correspond to the 11 Sefirot formulation of later schools. 113
Figure 17.3. The descent to below of the ten spaces with their spectrum number and their space-time dimensions indicated within each oval. The set of 10 spaces reside within a space of 1024 × 1024 dimensions (the Superverse) that is not displayed. 114
Figure 17.4. A 64 dimension 64-block has 32 complex dimensions. It is one layer in QUeST. The 32 = 10 Sefirot plus 22 channels correspond to 32 complex dimensions (64 real-valued dimensions) of a QUeST layer. 115
Figure 17.5. The 32 emanations—10 Sefirot and 22 channels (letters)—related to the fundamental representations groups of a QUeST layer.......... 116
Table G-6.1. The entities of existence. The Cosmos could have been viewed as a spirit if all forces between its parts are zero. In reality there are forces and so the Cosmos is "material" as well. 121
Figure 19.1. The ten spaces with their spectrum number and their space-time dimensions indicated within each oval as in Fig. 1.4. The number of spinor components for each fermion - antifermion pair that annihilates to produce the "next space down" is specified next to each arrow for the lower spaces. On the right are symbols for instances (particles) of spaces. 138
Figure 40.1. Feynman diagram for conventional and the nth diagram of a cloaked Two-Tier propagator. 148
Figure 50.1. A visualization of the set of General Coordinate transformations separated into real-valued General coordinate transformations (part I), complex transformations that satisfy $\Lambda(\omega, u)^T G \Lambda(\omega, u) = G$ (part II), and complex transformations that do not satisfy $\Lambda(\omega, u)^T G \Lambda(\omega, u) = G$ (part III). Part I and part II combine in the limit of flat space-time to form the Complex Lorentz group. Parts II and III elements form a U(4) group that we call the General Relativistic Reality group.......... 175
Figure C.1. The 32 complex quaternion dimension QUeST array. This array is the 32 × 8 array of •'s. 183
Table C.1. Map between fundamental representations and their dimensions. 184
Figure C.2. The 32 complex quaternion dimension QUeST array subdivided into 4 layers of 8 rows..Each layer will be seen to map to a block of fundamental group representations as shown in Figs. C.3 and C.4. 185

Figure C.3. The four layers of QUeST internal symmetry groups (and space-time) for 32 dimension complex quaternion space. Note: each row has an 8 • complex quaternion. Note the left column of blocks combine to specify a 4 dimension complex quaternion space-time. Note each layer requires 64 dimensions. *The U(2) Dark group is broken to U(1)⊗U(1) in chapter 2 to conform to the breakdown to 4 × 4 blocks suggested by comparison to the spinor structure.* 186
Figure C.4. The internal symmetry groups for one QUeST layer of the 4 layers in the 32 × 8 dimension array format. The two large blocks are each 5 complex dimension (10 real dimension) representations of SU(2)⊗U(1)⊗SU(3). The U(2) group (badly broken) supports transformations (rotations) between Normal and Dark matter. 187
Figure C.5. Fermion particle spectrum and partial examples of the pattern of mass mixing of the Generation group and of the Layer group. Unshaded parts are the known fermions with an additional, as yet not found, 4th generation. The lines on the left side (only shown for one layer) display the Generation mixing within each layer. The Generation mixing occurs within each layer using a separate Generation group for each layer. The lines on the right side show Layer group mixing (for Dark matter) with the mixing among all four layers for each of the four generations individually. There are four Layer groups for Normal matter and four Layer groups for Dark matter.. There are 256 fundamental fermions. QUeST and UST have the same fermion spectrum........ 188
Figure C.6. Fundamental fermions have a 1:1 correspondence with QUeST dimensions. Note the number of dimensions in each row is 8 – the number of dimensions in a complex quaternion. Correspondingly the number of fermions in each row is 8 – a suggestive similarity. Each layer has four normal fermion generations and four Dark fermion generations. Each dot (pebble) represents a dimension in the left part of the figure and a fermion in the right part. 189
Figure C.7. The 16 × 16 array of QUeST dimensions. 190
Figure C.8 The 16 × 16 form of QUeST array has two "layers", each of which is composed of two layers of the 32 × 16 four layer figure of Fig. C.3. Note each of these "layers" has 128 dimensions. 191
Figure C.9. Set of 16 dimension blocks for the first "layer" (of two "layers") of the 16 × 16 array. Each "dashed" block (regardless of its apparent size) is a 4 × 4 = 16 array of dimensions. This set of 8 blocks contains the 8×16 = 128 dimensions of "layer" 1. "Layer" 2 is similar. The Dark U(2) groups supports transformations (rotations) between the types of matter: Normal and Dark1; and Dark2 and Dark3. 192
Figure C.10. The two "layers" of 4 × 4 dimension subblocks of the 16 × 16 dimension array. 193
Figure C.11. Spectrum of the generations of fermions of QUeST for the 16 × 16 dimension array representation. Each fermion is represented by a •. Quark triplets are represented by three •'s. Note there are 256 fundamental fermions. 194

Figure C.12. Block form of a 16 × 16 QUeST fermion array with each block row corresponding to one layer. Each block contains four generations of fermions. The result is 4 × 4 blocks. The label e q-up indicates a charged lepton – up-type quark pair, ν q-down indicates a neutral lepton – down-type quark pair, and so on. Note the blocks can be reaaranged into a 32 × 8 form without physical consequences at this level of discussion since the right two columns and the lowest two rows are all Dark at present. 194
Figure C.13. The 32 complex quaternion dimension QUeST array partitioned down to a real spwce-time. The partition labeled "1" reduces the array to 32 quaternion dimensions after discarding the right columns. The partition labeled "2" reduces the array to 32 complex dimensions similarly. The partition labeled "3" reduces the array to 32 real-valued dimensions in the leftmost column.. 195
Figure C.14. Spectrum of a generation of fermions. Note this is only one layer of fermions. Two leptons and 6 quarks in each row of Normal matter. Similarly for Dark matter. 196
Figure E.1. The 64 complex octonion dimension UTMOST array. This is the 64 × 16 array of •'s. It has 1024 dimensions. 199
Figure E.2. The UTMOST array with 32 × 32 dimensions for a 32 quaternion octonion dimension space. 200
Table E.1. Map between fundamental representations and their dimensions 200
Figure E.3. The internal symmetry groups of *one layer* (consisting of 8 rows in Fig. E.2) of the four layers of 32 × 32 dimension UTMOST. The other three layers are copies of the this layer. Note the Dark U(4) groups. One U(4) "rotates" among Normal, Dark1, Dark2, and Dark3. The other U(4) "rotates" among Dark4, Dark5, Dark6, and Dark7. The U(4) group may be broken to U(1)⊗U(1)⊗U(1)⊗U(1) to obtain 4 × 4 blocks.... 201
Figure E.4 The *first* of the four layers of UTMOST dimensions with boxes around sets of dimensions for fundamental group representations. The U(4) Dark groups have been broken into U(2)⊗U(2) factors. The U(4) group may be further broken to U(1)⊗U(1)⊗U(1)⊗U(1) to obtain 4 × 4 blocks. 202
Figure E.5. Four layers (each in two rows) in the 32 × 32 dimension UTMOST array composed of 4 × 4 blocks, which are within the four block 8 × 8 sections for each pair: Normal+Dark1, Dark2+Dark3, Dark4+Dark5 and Dark6+Dark7. In total they form the 32 × 32 = 1024 UTMOST dimension array 203
Figure E.6. Spectrum of UTMOST fermions in a 16×64 format. Each fermion is represented by a •..Each set of eight •.'s represents a charged lepton, a neutral lepton, three up-type quarks, and three down-type quarks. There are eight sets of four species in four generations which are in turn in 4 layers. There are 1024 fundamental fermions taking account of quark triplets. 204
Figure E.7. Block form of the 32 × 32 UTMOST fermion array with each row corresponding to *half of an UTMOST layer*. Thus 8 × ½ = 4 layers results. Each block

contains four generations of fermions. The result is sixty-four 4 × 4 blocks. The label e q-up indicates a charged lepton – up-type quark pair, ν q-down indicates a neutral lepton – down-type quark pair, and so on. The form displayed here may explain why generations come in fours. 205
Figure E.8. The partition of the 32 × 32 dimension UTMOST array into MOST subspaces The size of each subspace is 32×16 = 512 dimensions. 206
Figure E.9. Partition of spectrum of UTMOST fermions in a 16×64 format. Each fermion is represented by a •. Including each quark. Each set of eight •.'s represents a charged lepton, a neutral lepton, three up-type quarks, and three down-type quarks. There are eight sets of four species in four generations which are in turn in 4 layers. There are 512 fundamental fermions in each subspace taking account of quark triplets. Note: Quark singlets won't do; triplets are required. 207
Figure E.10. Partition of block form of the 32 × 32 UTMOST fermion array with each row corresponding to *half of an UTMOST layer*. Thus 8 × ½ = 4 layers results. Each block contains four generations of fermions. The result is sixty-four 4 × 4 blocks. The label e q-up indicates a charged lepton – up-type quark pair, ν q-down indicates a neutral lepton – down-type quark pair, and so on. 208

Introduction

This volume is the Second Edition of *From Octonion Cosmology to the Unified SuperStandard Theory of Particles*. It presents a more detailed, complete Cosmology based on a spectrum of ten octonion spaces, a Superverse, and three spaces of functionals. Octonion Cosmology is a truly fundamental physical theory that "explains" space-time, internal symmetry groups, and elementary particles, and their associated, detailed phenomena. It provides a justification for the Unified SuperStandard Theory that the author previously derived from Logic considerations over the past twenty years (based on the author's extension of Quantum Field Theory presented in papers in the 1970s.) There is a panoramic range from the most fundamental physical levels to current elementary particle physics experiments.

The Second Edition covers some additional topics not presented in the First Edition:

An Octonion Spaces Computation Model
More Information on Strings, Bags, Data Spaces, Minispaces, Higgs Bosons
More Parallels to *Sefer Yetzirah* Cosmology, Linguistic Theory & God-View
Sefer Yetzirah Hints of Lost Ancient Physics
The Meaning of the 231 Gates "in the Wall" to Space-Time
A God-View Based on Logic Using Monads
Possible Cascading Quantum Probability Tweaks to Manage History
Thoughts on Life After Death Through Monads?

The past 100 years have witnessed major growth in our understanding of elementary particles and Physical Reality. At this point in time it is appropriate to create a synthesis of the cumulative acquired physical knowledge. This book combines the work of the author over the past fifty years to develop a synthesis. The result is a complete theory of Elementary Particles and Gravitation centered on a generalization of space that combines internal symmetries and space-time in a relatively simple theory built ultimately on a space, called Maxiverse, with a ten dimension space-time.

Maxiverse space supports a primitive type of fermion called an urfermion. This fermion and its antiparticle annihilate to produce the 1024 dimension Megaverse instance (particle). A Megaverse in turn supports "seed" fermions. A seed fermion and its antiparticle can annihilate to generate a universe particle containing a QUeST (Quaternion Unified SuperStandard Theory) universe with 256 dimensions.

A QUeST universe contains the internal symmetries of the UST (Unified SuperStandard Theory) and a four octonion dimension space-time. UST contains The Standard Model as well as additional reasonable symmetries.

The result is a complete theory! Blaha (2020c) and its predecessors describe UST in detail. A series of books this year by the author describes the deeper basis of

UST in the QUeST, BQuEST, UTMOST, and BMOST formulations of Octonion Cosmology.

Octonion Cosmology consists of a unification of ten closely connected octonion spaces into a framework that combines Megaverse space, universe space, and miniverse space.

The Quaternion Unified SuperStandard Theory (QUeST) is a theory for universes. It implies the Unified SuperStandard Theory (UST) of the author to the author's initial surprise. The Megaverse theory UTMOST (Megaverse Octonion SuperStandard Theory) is also based on Octonion Cosmology. Both theories are based on deeper theories BQUeST and BMOST described in this book.

This book provides a detailed derivation of QUeST from a Megaverse fermion theory and describes the origin of universes as particles. It also provides a detailed derivation of UTMOST from a ten dimension Maxiverse and provides details of the origin of a Megaverse particle. It reduces the origin of universes and of the Megaverse to the simplest possible basis. An important aspect of these derivations is the relation of universes to the Megaverse. The one fermion seed for universes is in the Megaverse; the one fermion seed for the Megaverse is in a ten dimension space-time that may be equivalent to a SuperSymmetry space.

This book also proposes a clear multilevel sequential pattern of symmetry breaking of QUeST down to the level of SU(3) and SU(2)⊗U(1) symmetries. It also considers the detailed derivation of UST from QUeST, and provides an .Overview of QUeST and its UST sector features. An Overview of Megaverse UTMOST features is also presented.

This book completes the panorama of hypercomplex features for QUeST universes and the UTMOST Megaverse. The result is a complete theoretical framework for Elementary Particles and Gravitation.

In the interests of completeness the book contains appendices detailing aspects of the Unified SuperStandard Theory (UST), QUeST, and UTMOST for use in derivations. A particularly important appendix, Chapter 17, describes the close analogy of Octonion Cosmology and the Cosmology implicit in the *Sefer Yetzirah*. The correspondence leads to the hope of an eventual convergence of Physics, Philosophy and Religion.

The combined theory of UST, QUeST, UTMOST, and features of the other Octonion spaces form a great and exciting intertwined overall theory of everything—in some sense the "Philosopher's Stone" of Physics.

0. Hypercomplex Spaces - Extension of the Unified SuperStandard Theory (UST) to Higher Dimensions

0.1 Suggestions of a Deeper Basis for UST

The UST provides a complete theory of elementary particles and Gravitation in 3 + 1 dimensions (real-valued coordinates). It is presented in detail in Blaha (2020c). In the study of this theory extending back 7+ years to Blaha (2012a) and earlier books, the author found a close analogy between the subgroups of the Complex Lorentz Group and the factors of the Standard Model internal symmetry: SU(2)⊗U(1)⊗SU(3).

This seeming coincidence raised the question that the analogy was based on a deeper relation embodied in a larger space. This larger space would have a subspace devoted to space-time and a subspace for internal symmetry groups. The subspaces would be orthogonal.

Pursuing this concept led the author, in the fall, 2019, to develop a formulation of larger spaces based on hypercomplex coordinates: quaternions and octonions. This development reached fruition in Blaha (2020a) through (2020k). *The author found an almost complete correspondence (which was easily made complete) between the Unified SuperStandard Theory based on Logic and the QUeST theory based on hypercomplex numbers.*

0.2 Hypercomplex Numbers

The key concept in the search for a deeper basis for UST was the following progression of coordinate system formulations:

A. The Standard Model was based on real-valued coordinates.

B. UST is based on real-valued coordinates partly made complex:

B1. Real-valued coordinates were replaced by slightly complex coordinates with an imaginary part consisting of a massless second quantized vector field that gave the usual perturbation theory results at low energy and \eliminated divergences at very high energy.

B2. Quarks and Strong Interaction gauge fields were truly complex (with certain restrictions) due to Complex Lorentz Group boosts.

C. Having progressed from real-valued coordinates to complex-valued coordinates in parts A and B, we considered hypercomplex-valued coordinates. The author found that hypercomplex space was important for the correspondence with UST. This was discussed in Blaha (2020c). The choice of coordinates evolved culminating in Blaha (2020k) and this book. The coordinate systems of choice now are listed in Figure 1.1 below.

The result of this chain of progression was a theory of universes (called QUeST) that directly becomes UST upon restriction to real-valued coordinates. We also found that a theory called UTMOST corresponded to a Megaverse (Multiverse), within which QUeST universes could exist.

Thus there are three closely related theories UST, QUeST, and UTMOST. There are also 9 other spaces of interest. See Fig. 1.4.

0.3 Mathematical Picture Language

As we will see later we need only consider the allocation of dimensions to space-time, and to the various internal symmetry groups in hypercomplex spaces. The detailed mathematics of quaternions and octonions is not directly relevant for the internal symmetry subspaces. *We only use hypercomplex coordinates to specify dimensions in this and earlier books.* The algebras of hypercomplex coordinates may be relevant in the definition of quantum fields and the evaluation of perturbation theory calculations.

Since the allocation of dimensions to fundamental representations of groups is of paramount importance to establish the basis of UST, we simply used the symbol • to represent a dimension. The many dimensions within hypercomplex coordinates are thus best represented by patterns of •'s. The separation of the dimensions (•'s) into fundamental representations is easily accomplished.

We have a simple Mathematical Picture Language for the representation of dimensions for the separations that would require complex transformations otherwise as indicated in the footnote below.

0.4 Number Describes Everything

Pythagoras developed a language of •'s for computation. He called the •'s *psiphi*, meaning pebbles in Classical Greek. He developed a set of psiphi diagrams for numbers. The perfect number 10 was represented by

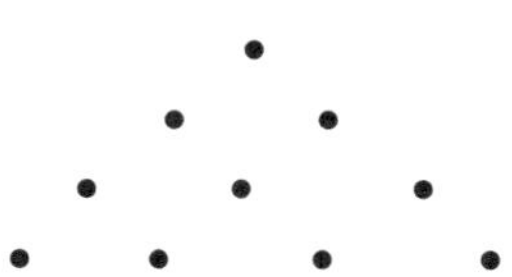

Figure 0.1. The Pythagorean tetrackys, a psiphi representation of 10.

Pythagoreans considered ten to be the perfect number because it embodies 1, 2, 3, and 4. Remarkably perhaps, the groups U(1), U(2), SU(3), and U(4) are the internal symmetry groups of UST and the Standard Model.

Pythagoras' great achievements and his innate sense of the nature of Reality led to the proposition:

The essence of all things is number. (0.1)

Our theories directly reflect its truth.

From the total symmetry of each internal symmetry subspace of each of the ten spaces in Fig. 1.4,, and with the use of the number of dimensions of fundamental representations of internal symmetry groups, we are led directly to the internal symmetry groups, the fermion spectrum, the vector boson spectrum and Gravitation of UST.

We find "blocks" of fundamental representations of the groups known from the Standard Model, and UST, appearing in the octonion-based theories:

$$SU(2)\otimes U(1)\otimes SU(3)\otimes SU(2)\otimes U(1)\otimes SU(3) \tag{0.2}$$

where the second set of SU(2)⊗U(1)⊗SU(3) factors are symmetries of Dark matter in UST. The total fundamental representations' dimension of these factors is 10, the Pythagorean perfect number—an interesting coincidence. It suggests the Pythagorean summation of Reality should be:

Number is the Essence of Eternity (0.3)

since one can hope the theories, UST, QUeST, and UTMOST, if correct, are "eternal" being all based on Number.

0.5 Axioms for the Unified SuperStandard Theories

The Unified SuperStandard Theory is derived, in a Euclidean manner, from a set of Axioms. We list a set of Axioms for the Unified SuperStandard Theory for our 3 + 1 dimension universe, for the 3 + 1 complex quaternion QUeST universe, and for the 7 + 1 octonion UTMOST Megaverse.

Remarkably the space of type 6 in Fig. 1.4 corresponds directly with UST (with one minor change noted later). We present the set of Axioms modified slightly from those of Blaha (2020c).

UST/UTMOST AXIOMS

1. Hypercomplex spaces are the basic spaces of our universe and the Megaverse.
2. Physical processes can execute in parallel.
3. Matter and energy are particulate.
4. Space--times are locally Lorentzian.
5. All calculations are finite.
6. Particle theory can be defined in any curved space-time.
7. Each particle has a wave function determined by a functional inner product defining the particle state. The functionals form a set without a distance measure.

0.6 General Implications of the Axioms

In this section we describe some of the implications of each of the axioms.

1. Hypercomplex spaces are the basic spaces of our universe (type 6) and the Megaverse (type 5). See Fig. 1.4.

The factorization into space-times and an internal symmetry space must be a form of spontaneous symmetry breaking of yet unknown origin. It appears to be related to a breakdown of the vacuum.

2. Physical processes can execute in parallel.

Physical processes are known to be able to execute in parallel at any distance of separation. As Fant has shown parallel execution requires a minimal number of dimensions: 4. Consequently the dimension of space-time must be 4 or greater. The space-time of QUeST is 4-dimensional allowing parallel process execution.

The space-time of UTMOST is 8-dimensional and also allows parallel process execution. The choice of eight dimensions is natural since it allows 4-dimensional universes within it. It also has a form that allows a clean formulation. Lastly, as will be seen later, it conforms to the pattern of interplay between Lorentz symmetry and internal symmetry found in the Unified SuperStandard Theory. This axiom leads to a view of the origin of the dimensions.

3. Matter and energy are particulate.

The most direct method of specifying a theory of matter and energy is through the Use of Quantum Field Theory. Thus Quantum Field Theory is implied.

4. Complex Space-times are locally Lorentzian.

A locally complex Lorentzian space-time leads to Complex General Relativity. In flat space-time Complex General Relativity becomes Complex Lorentz group. (In point of fact the Complex Poincaré group follows.)

5. All calculations are finite.

Given the need for Quantum Field Theory it becomes necessary to find a formulation that yields finite values for calculations in perturbation theory. The only approach that eliminates high energy divergences, and yet preserves the results found in perturbation theory calculations that agree with (primarily QED) experiments, is Two-Tier Quantum Field Theory. This is discussed in detail in earlier books starting in 2002. Thus only our Two-Tier formalism satisfies this axiom.

6. Particle theory can be defined in any curved space-time.

In the 1970s we developed a formalism that allows the definition of particle states in any space-time in such a way that its physical content is preserved when transformed to any coordinate system.[1] This PseudoQuantum Quantum Field Theory satisfies this axiom.

7. Each particle has a wave function determined by a functional inner product defining the particle state. The functionals form a set without a distance measure.

This axiom is satisfied by our formulation of quantum functionals in Blaha (2019f) and earlier books. Our formulation eliminates the superficial violation of the Theory of Relativity by "spooky" quantum entangled processes with parts separated by a physically "large" distance.

The seven axioms imply the Unified SuperStandard Theory and its deeper biquaternion and bioctonion hypercomplex formulations.

---

[1] S. Blaha, Il Nuovo Cimento **49A**, 35 (1979).

0.7 Hypercomplex Number Based Higher Dimensions

The Unified SuperStandard Theory was based on real-valued and complex-valued coordinates. This choice enabled us to understand the reason behind the four types of fermions found in nature: neutral fermions, charged fermions, up-type quarks, and down-type quarks. A study of Complex Lorentz group subgroups showed that they were similar to the factors of the Standard Model symmetry group. This similarity motivated this author to consider spaces that integrated space-time and internal symmetry dimensions.

In choosing a higher dimension space for a larger theory of elementary particles, the use of coordinate systems based on hypercomplex number systems seemed reasonable. The hope was, that just as complex coordinates led to a deeper understanding of the internal symmetries of the Standard Model, the use of hypercomplex coordinates might lead to a further understanding of the origin of both space-time and internal symmetries.

The use of quaternion and octonion coordinate systems was motivated by the nature of S matrix elements, which necessarily use the Dyson-Wick expansion of time ordered products in perturbation theory.[2] S matrix element calculations require the arithmetic rules of real numbers or complex numbers or quaternion numbers or octonion numbers for calculations.

The pattern of rising hypercomplexity is:

Real → Complex → Quaternion → Octonion

The Unified SuperStandard Theory took particle theory from real-valued coordinates to complex-valued coordinates. Hypercomplex coordinates take us to QUeST and UTMOST. They both use larger spaces to unite space-time symmetry and internal symmetry.

Since the requirement of parallel physical processes[3] made the minimal space-time dimension 4 and since the Megaverse must include universes as subspaces, we were led to a 4-dimensional space-time formulation for our universe and an 8-dimensional space-time formulation for the Megaverse. They were considered in Blaha (2020a), (2020b) and Blaha (2020c).

In Blaha (2020c) we considered these higher dimensional theories and found that the complex quaternion higher dimensional theory QUeST) leads directly to the 3 + 1 dimensional Unified SuperStandard Theory upon restriction of the theory to real-valued coordinates. Similarly the Megaverse UTMOST leads to a reasonable Megaverse theory upon restriction to a generalization of the Unified SuperStandard Theory in 7 + 1 real space-time dimensions.

---

[2] Time-ordered products are defined in quaternion perturbation theory in terms of a "time" defined as a quaternion inner product of a chosen direction in quaternion time, and the time quaternion. Quaternion time is thus single-valued. Similar comments apply to octonion perturbation theory. Potential infinities in higher dimension perturbation theory are eliminated using the author's Two Tier formulation of Quantum Field Theory. See Blaha (2005a).

[3] See Blaha (2020c).

In Blaha (2020k) and this volume we propose a spectrum of octonion-based spaces, which include a space for universes and for Megaverses (types 6 and 5 in the spectrum of spaces in Fig. 1.4 respectively.)

0.8 From Complex Coordinates to Hypercomplex Coordinates

In this section we show a natural generalization of coordinate systems to hypercomplex number systems. There are a number of ways to define hypercomplex coordinate systems. We extrapolate from the form of complex coordinates:

$$
\begin{aligned}
t &= t_1 + it_2 \\
x &= x_1 + ix_2 \\
y &= y_1 + iy_2 \\
z &= z_1 + iz_2 \\
&\cdots
\end{aligned}
$$

Note that the real part of each coordinate is the same type as the imaginary part—also a real number.

0.9 Quaternion Coordinate Systems

Based on this simple fact it seems natural to define a 3 + 1 dimensional space-time with quaternion coordinates as:

Time quaternion

$$t = (a + ib + jc + kd) \tag{0.4}$$

Spatial quaternions

$$
\begin{aligned}
x &= (a_x + ib_x + jc_x + kd_x) \\
y &= (a_y + ib_y + jc_y + kd_y) \\
z &= (a_z + ib_z + jc_z + kd_z)
\end{aligned}
$$

where a, b, c, d are real-valued numbers. The symbols i, j, and k are fundamental quaternion units. Quaternions are described in Appendix 0-A. A quaternion embodies four coordinates. Thus it is a 4-dimensional entity and we attribute 4 dimensions to each quaternion and 8 dimensions to each complex quaternion.

In the UST development (Blaha (2020c)) we saw that complex coordinates and the Complex Lorentz Group were needed to find the four types (species) of fundamental fermions[4] and the internal symmetry groups. Consequently we used complex quaternions[5] (biquaternions) to develop QUeST in Blaha (2020c), initially in a 7 + 1 complex quaternion space.

Our primary interest in hypercomplex spaces is their dimensions and how they can be transformed into a space-time part and an internal symmetry part. This purpose is best approached by simply displaying a Mathematical Picture Language

---

[4] Charged leptons, neutral leptons, up-type quarks, and down-type quarks.

[5] We changed thee space-time to four octonion coordinates in Blaha (2020k) and here.

representation using Pythagoras' *psiphi* (pebbles) diagrams where each dimension is represented by a • in Fig. 0.2.

The real-valued dimensions[6] of the 3 + 1 space-time of eq. 0.4 and Fig. 0.2 number 32.

Time

•••• ••••

Space

•••• ••••

•••• ••••

•••• ••••

Figure 0.2. Psiphi diagram of the dimensions of 3 + 1 complex quaternion space. Each row represents a complex quaternion. Note the coordinates of eq. 0.4 map directly to dimensions represented by •'s.

0.10 Octonion Coordinate Systems

The use of octonion coordinate systems is consistent with the calculation of S matrix elements using the Dyson-Wick expansion of time ordered products in perturbation theory.[7] S matrix element calculations require the arithmetic rules of real numbers or complex numbers or quaternion numbers or octonion numbers for calculations. Octonions can be used to define a coordinate system. An octonion has 8 coordinates and thus it has dimension 8. See Appendix 0-B for a brief description of octonions.

We can define a 3 + 1 dimensional real-valued octonion space as:

Time Octonion

$$t = a + ib + jc + kd + h'a' + i'b' + j'c' + k'd' \tag{0.5}$$

Spatial Octonions

$$x = a_x + ib_x + jc_x + kd_x + h'a'_x + i'b_x' + j'c_x' + k'd_x'$$
$$y = a_y + ib_y + jc_y + kd_y + h'a'_y + i'b_y' + j'c_y' + k'd_y'$$
$$z = a_z + ib_z + jc_z + kd_z + h'a'_z + i'b_z' + j'c_z' + k'd_z'$$

where a, b, c, d, a′, b′, c′, d′ are *real-valued* numbers. The symbols i, j, k, h′, i′, j′ and k′ are fundamental octonion units.

In the UTMOST development that we considered previously, we saw that complex octonions were needed initially in a 7+ 1 dimensions complex octonion space-time. The 7 + 1 dimensions complex octonion space-time coordinates are:

---

6 Real-valued coordinates are said to have real-valued dimensions. Complex coordinates have complex-vslued dimensions.

7 Time-ordered products are defined in octonion perturbation theory in terms of a "time" defined as a octonion inner product of a chosen direction in octonion time, and the time octonion. Octonion time is thus single-valued. Potential infinities in higher dimension perturbation theory are eliminated using the author's Two Tier formulation of Quantum Field Theory. See Blaha (2005a).

Time Octonion

$$t = a + ib + jc + kd + h'a' + i'b' + j'c' + k'd' \quad (0.6)$$

Spatial Octonions

$$x = a_x + ib_x + jc_x + kd_x + h'a'_x + i'b_x' + j'c_x' + k'd_x'$$
$$y = a_y + ib_y + jc_y + kd_y + h'a'_y + i'b_y' + j'c_y' + k'd_y'$$
$$z = a_z + ib_z + jc_z + kd_z + h'a'_z + i'b_z' + j'c_z' + k'd_z'$$
$$x1 = a_{x1} + ib_{x1} + jc_{x1} + kd_{x1} + h'a'_{x1} + i'b_{x1}' + j'c_{x1}' + k'd_{x1}'$$
$$y1 = a_{y1} + ib_{y1} + jc_{y1} + kd_{y1} + h'a'_{y1} + i'b_{y1}' + j'c_{y1}' + k'd_{y1}'$$
$$z1 = a_{z1} + ib_{z1} + jc_{z1} + kd_{z1} + h'a'_{z1} + i'b_{z1}' + j'c_{z1}' + k'd_{z1}'$$
$$w1 = a_{w1} + ib_{w1} + jc_{w1} + kd_{w1} + h'a'_{w1} + i'b_{w1} + j'c_{w1}' + k'd_{w1}'$$

where all coefficients: a, b, c, d, a′, b′, c′, d′, and a_i, b_i, c_i, d_i, a'_i, b'_i, c'_i, d'_i for i = x, y, z, w, x1, y1, z1, w1 are ***complex-valued*** numbers. Each complex octonion embodies 16 dimensions. See Fig. 0.3 for the psiphi representation of dimensions in eq. 0.6.

0.11 "Composite" Octonion Coordinate Systems

A quaternion has 4 dimensions associated with it. An octonion has 8 dimensions associated with it. An octonion can be viewed as a pair of quaternions. However as we shall see in the table of Octonion Cosmology spaces (Fig. 1.4) using octonions gives a cleaner regular pattern of the spectrum of spaces.

Fig. 1.4 uses a form of "composite" octonion generalization that does not appear to have been used previously. In this section we will illustrate the forms of composite octonions with some examples.

Example 1:

Octonion octonion specifies an "octonion of octonions" of number or dimension 8×8 = 64.

Example 2:

Complex octonion octonion specifies a "quaternion of octonion of octonions" of number or dimension 4×8×8 = 256.

Example 3:

Complex Octonion Octonion Octonion specifies "complex octonion of octonion of octonions" of number or dimension 2×8×8×8 = 1024.

The composite octonion representation is used in Fig. 1.4 to show the regular pattern of space dimensions in Octonion Cosmology.[8]

[8] One could reexpress the table in Fig. 1.1 in terms of quaternions. That would result in a more cumbersome table without the clean ordering of the composite octonion display

Appendix 0-A. Quaternion Features

Quaternions and octonions are hypercomplex numbers with special properties that make them similar to complex numbers.[9] Quaternions and octonions are both normed division algebras over the reals (*hypercomplex* number systems) with salutary properties for quantitative studies in quantum field theory and perturbation theory. Some of their new features are listed on the cover page.

Quaternions have significant properties that distinguish them:

1 .They are associative.

2. They are one of the two finite dimensional division rings having the real numbers as a proper subring. (The other is octonions—considered in chapter 8.)

3. They are non-commutative. (This is not a roadblock for quantum field theory which is also non-commutative in general.)

These features support the development of physics theories.[10]

0-A.1 Some Basic Quaternion Features

A quaternion is a 4-tuple of real numbers. A complex quaternion is a 4-tuple of complex numbers:

$$x = a + bi + jc + kd \quad = a + \mathbf{v} \tag{0-A.1}$$

where a, b, c, d are real or complex numbers, and **v** is a 3-vector. The symbols i, j, and k are fundamental quaternion units. A quaternion norm is defined by

$$||\mathbf{x}|| = \text{sqrt}(aa^* + bb^* + cc^* + dd^*) \tag{0-A.2}$$

and the norm of **v** is

$$||\mathbf{v}|| = \text{sqrt}(bb^* + cc^* + dd^*) \tag{0-A.3}$$

An important identity is

$$e^x = e^a \left(\cos(||\mathbf{v}||) + \mathbf{v}/||\mathbf{v}|| \sin(||\mathbf{v}||)\right)s \tag{0-A.4}$$

.It is used to define boosts in quaternion space.

---

[9] Much of this appendix appears in Blaha (2020a) and (2020b).

[10] There is an extensive literature on quaternions starting with the original work of Hamilton. Some recent, relevant papers are: S. L. Adler, "Generalized Quantum Dynamics", IASSNS –HEP-93/32 (1993); S. De Leo, arXiv:hep-th/9506179 (1995); Rolf Dahm, arXiv:hep-th/9601207 (1996); S. De Leo, arXiv:hep-th/9508011 (1995); S. L. Adler, arXiv:hep-th/9607008 (1996) and references therein.

Appendix 0-B. Some Octonion Features

Octonions have significant properties that enable them to be used in a quantum field theory development:

1. An octonion is an 8-tuple of real numbers. A complex octonion is an 8-tuple of complex numbers.
2. They are nonassociative.
3. They are one of the two finite dimensional division rings having the real numbers as a proper subring. (The other is quaternions—considered in chapter 5.)
4. They are non-commutative. (This is not a roadblock for quantum field theory which is also non-commutative in general.)

These features support the development of physics theories.

We can represent a complex octonion (bioctonion) b as

$$b = b_{real} + Ib_{imaginary}$$

where b_{real} and $b_{imaginary}$ are real-valued octonions. We can also represent a complex octonion as an octonion with complex coordinates as in eq. 0.8.

1. The Spectrum of 10 Octonion Spaces

The search for deeper roots of the Standard Model of Elementary Particles led us to derive a Unified SuperStandard Theory (UST) with the Standard Model as a subset. The derivation of the UST was based on an approach similar to that of Geometry. It started with a set of axioms and proceeds systematically to derive features. The features, that were found, were a superset of the Standard Model including a Dark Matter sector, four layers of particles and interactions, and additional interactions.

A significant similarity[11] between the symmetries of the Standard Model and subgroups of the Lorentz group led us to explore the possibility of a further unification of internal symmetries and space-time symmetries within the framework of a higher dimension theory.

Our search (in books in 2020—See References) led us to successfully base UST on Quaternion[12] Unified SuperStandard Theory (QUeST) with a remarkable match between the internal symmetries, interactions, particle spectrum, and space-time symmetry of both theories.

We described[13] a Cosmology based on a spectrum of octonion spaces that included interlocked universe and Megaverse instances (particles) of two octonion spaces of the 10 octonion spaces. This Cosmology appears to be at the deepest level of physical reality. In this chapter we describe the features of its ten spaces in greater detail as well as an enclosing *Superverse* octonion space of 1,000,000+ dimensions. We also describe additional details of the spaces and their interrelations in the chapters that follow.

1.1 Fundamental Axioms

We start with two fundamental axioms with the goal of creating a unified theory of internal symmetries and space-time, *without* difficulties with well-known NO GO theorems. We choose to deal with the most general form of number: octonions because their algebra limitations: non-commutativity and non-associativity are analogous to similar features in Quantum Field Theory:

$$\psi_1\psi_2 \neq \psi_2\psi_1$$
$$(\psi_1\psi_2)\psi_3 \neq \psi_1(\psi_2\psi_3)$$

where ψ_1, ψ_2 and ψ_3 are quantum fields. Thus a perturbation theory built on octonions is feasible.

---

[11] Blaha (2018e) and (2020c) as well as other books by the author.

[12] In Blaha (2020l) we established QUeST in an octonion framework.

[13] See Blaha (2020l).

In the study of octonion spaces in this book we do ***not*** *use octonion algebra. We treat octonions as determining the specific dimensions of higher dimension octonion spaces. We will initially specify the dimensions of octonion spaces based on certain axioms below. Subsequently we will map the UST, and related internal symmetry groups and space-times, to octonion spaces.*

We assume

1. An octet (octonion) of octonion spaces.

2. The spaces' dimensions progressively grow by factors of four from space to space with the smallest space initially having the dimensions of eight octonion coordinates where each coordinate is an octonion. Thus the number of dimensions in the smallest space is $8 \cdot 8 = 64$.

3. Each space has a square array of dimensions.

We number the components of the spaces octet from 0 through 7 and call each component a *level*.[14] The effect of axiom 2 is to have the total number of dimensions of level j to be four times that of level j + 1 for j = 0, 1, … 6.

Axiom 3 follows naturally from the creation rules for octonion space instances from fermion-antifermion annihilation as described in chapters 3, 7, and 10. The initial square array of a space can be reexpressed (before connecting the space to Physics) as a rectangular array as we do later. The eventual shape of the array may be determined by the masses of the fundamental fermions. One form for UST has four layers with each layer having four generations of fermions. Each layer has a matching "Normal" fermions part and Dark fermions part. The average of the fermion masses of each layer grows layer by layer. This form is discussed in chapter 2.

In units of octonions the number of octonions *per coordinate* for levels 7 through 0 is 1, 2, 4, 8, 16, 32, 64, and 128. For level j the number of octonions in a coordinate is 2^{7-j} and the number of coordinates is 2^{7-j}. The total number of dimensions is $2^{14-2j+6} = 2^{20-2j}$ since an octonion has 2^3 dimensions within it.

For example, each level 0 coordinate has $128 \cdot 8 = 1024$ elements, and level 0 has 1024 coordinates, with the result that the level 0 space has 1,048,576 dimensions. The above axioms imply Fig. 1.4. We specify the coordinate type and number of coordinates in terms of octonion multiples as in section 0.11.

[14] The numbering is similar to that of Dante's *Inferno*. The numbers begin at 0 and increase with descending levels. The numbering becomes more acceptable as we consider the roles of the space levels later.

OCTONION SPACES LEVELS

Level Number

Level	Coordinate Type	Number of Coordinates	Number of Dimensions
0	Complex Octonion Octonion Octonion (2^7)	Complex Octonion Octonion Octonion	1024×1024
1	Octonion Octonion Octonion (2^6)	Octonion Octonion Octonion	512×512
2	Quaternion Octonion Octonion (2^5)	Quaternion Octonion Octonion	256×256
3	Complex Octonion Octonion (2^4)	Complex Octonion Octonion	128×128
4	Octonion Octonion (2^3)	Octonion Octonion	64×64
5	Quaternion Octonion[15] (2^2)	Quaternion Octonion	32×32
6	Complex Octonion[16] (2^1)	Complex Octonion	16×16
7	Octonion (2^0)	Octonion	8×8

Figure 1.1. The levels and dimensions of the octonion space octet. Note the lowest level is octonionic as it should be.

1.2 Instances of the Octonion Spaces

The spaces defined in Fig. 1.1 can have instances[17] (particles) created. Since we anticipate the existence of a space for universes and for Megaverses (Multiverses), we see a need to create universes instances (particles) within Megaverse (particles.) In chapters 3, 7, and 10 we describe particle creation mechanisms based on the annihilation of a fermion-antifermion pair in a parent instance. Thus an instance of space 4 generates an instance of space 5 (a Megaverse) through the annihilation of a fermion-antifermion pair. An instance of space 5 generates an instance of space 6 (a universe) through the annihilation of a fermion-antifermion pair. And so on. Later we will see that the features of the created instance are determined by the structure of the annihilating fermion-antifermion pair.

1.3 Nesting of Octonion Spaces

The spaces in Fig. 1.1 can be viewed as nested:

Space 0 has four copies of space 1
Space 1 has four copies of space 2
Space 2 has four copies of space 3
Space 3 has four copies of space 4
Space 4 has four copies of space 5
Space 5 has four copies of space 6
Space 6 has four copies of space 7

Figure 1.2. Nesting of octonion spaces.

---

15 In our earlier books in 2020 we also designated this 1024 dimension space (5) as 64 complex octonion space.

16 In our earlier books in 2020 we also designated this 256 dimension space(6) as 32 complex quaternion space.

17 Appendix D contains evidence for the particle nature of universes.

1.4 Match of Level 6 Space with UST and Level 5 with UTMOST

The eight spaces of the octet have dimensions but are otherwise unspecified at this point. We now identify the level 6 space instance with our universe's UST. We note that level 6 has 256 dimensions. An examination of the dimensions of the internal symmetry and space-time of the UST gives 256 dimensions as well if the four dimensions of UST are generalized to octonion dimensions. Chapter 5 shows this match in detail. Therefore we take level 6 space to be the space of our universe.

Based on this choice we can identify an instance of space 5 as our Megaverse. (Other Megaverse instances are possible.) We can further identify a space 4 instance as the "parent" of space 5 instances. We will assume (and show later in chapter 7) that there is only one instance of space 4. We call that instance the Maxiverse.

1.5 Octet Space 7

Octet space 7 presents a problem in that we cannot derive it as following from a fermion-antifermion annihilation in level 6. Chapter 12 shows that such a process generates an instance of a $4 \times 4 = 16$ dimension space. That space in turn can generate another $4 \times 4 = 16$ dimension space, which can in turn generate yet another $4 \times 4 = 16$ dimension space, and another. We can truncate the generation of such spaces to yield four spaces in total. The four spaces, if combined, give the $8 \times 8 = 64$ dimension space of Fig. 1.1.

The resulting pattern of fermion-antifermion annihilation generations appears in Fig. 1.3.

1.6 The Spectrum of Physical Octonion Spaces

The result is a set of 10 spaces plus the aggregate space labeled 0. Thus there is a set of 11 conceptually interrelated spaces built on octonions. These spaces are listed in Fig. 1.4. Further types of octonion spaces can be defined. But they do not appear to be necessary, and their connection to the 11 spaces can be eliminated (truncated) by the specifications of the Complex Octonion Octonion[18] space (space 3) and the Quaternion Minispace space 10. We describe the limitation to 11 spaces in detail in later chapters.

[18] This phrase, and similar phrases, are not typos. The phrase "Octonion Octonion Octonion" means 8 octonion octonions, which means 64 octonions, which means a 512-tvector for a coordinate. This coordinate has 512 real values assembled in a 512-vector. Similar comments apply to other entries in Fig. 1.4. See chapter 0.

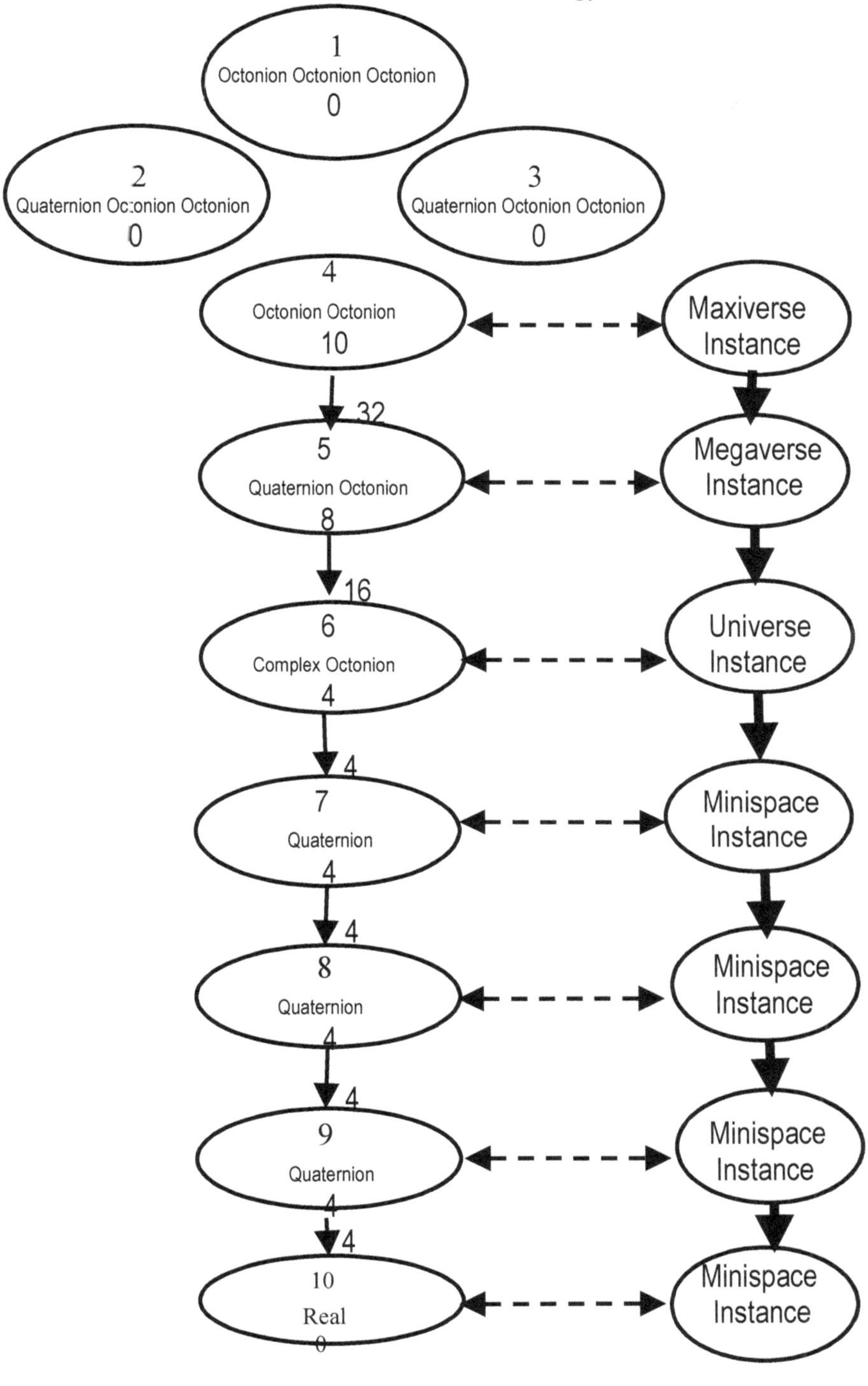

Figure 1.3. The ten spaces with their spectrum number and their space-time dimensions indicated within each oval as in Fig. 1.4. The number of spinor components for each fermion - antifermion pair that annihilates to produce the

"next space down" is specified next to each arrow for the lower spaces. On the right are symbols for instances (particles) of spaces.

OCTONION SPACES SPECTRUM

Spectrum Number

Spectrum Number	Coordinate Type	Number of Coordinates	Dimension Array Size	Space-Time Dimensions
	Superverse Space			
0	Complex Octonion Octonion Octonion (1024)	Complex Octonion Octonion Octonion	1024 × 1024	0
	Spaceless[19]			
1	Octonion Octonion Octonion (512)	Octonion Octonion Octonion	512 × 512	0
2	Quaternion Octonion Octonion (256)	Quaternion Octonion Octonion	256 × 256	0
3	Complex Octonion Octonion (128)	Complex Octonion Octonion	128 × 128	0
	Cosmology[20]			
4	Octonion Octonion (64) Maxiverse Space	Octonion Octonion	64 × 64	10 quaternion octonion
5	Quaternion Octonion[21] (32) Megaverses Space	Quaternion Octonion	32 × 32	8 complex octonion
6	Complex Octonion[22] (16) Universe Space	Complex Octonion	16 × 16	4 octonion
	Minispaces[23]			
7	Quaternion (4)	Quaternion	4 × 4	4 Real
8	Real (4)	Real (4)	4 × 4	4 Real
9	Real (4)	Real (4)	4 × 4	4 Real
10	Real (4)	Real (4)	4 × 4	0

Figure 1.4. The spectrum of the Superverse and the ten octonion spaces. The spaces are numbered from 0 through 10. The numbers in parentheses in column 2 are the number of dimensions in each coordinate. The items in column 3 are the number of rows of dimensions (1024, 512, 256, 128, 64, 32, 16, 4, 4, 4,4).

1.3 Characteristics and Roles of the Spaces

The spaces of Octonion Cosmology have a variety of features that ultimately lead to the Unified SuperStandard Theory (UST). We describe the relationships in detail. UST bears direct comparison to experiment and accommodates the known features of The Standard Model of Elementary Particles.

The spectrum of spaces is also of interest in its own right.[24]

---

[19] Spaceless spaces are spaces without a space-time.

[20] Cosmological spaces are those, which are directly related to physics from megaverses and universes to elementary particles.

[21] In our earlier books in 2020 we also designated this 1024 dimension space (5) as 64 complex octonion space.

[22] In our earlier books in 2020 we also designated this 256 dimension space(6) as 32 complex quaternion space.

[23] A Minispace is a subspace of a universe space.

1.3.0 Superverse Space

The Superverse is a space with no space-time coordinates within it. It is formless since there is no dynamics and thus has no symmetry breaking. It contains spaces 1 through 10 as subspaces. See Fig. 1.6 and chapter 13.

1.3.1 Spaceless Spaces (1 – 3)

The three spaceless spaces of Fig. 1.4 are each defined to have dimensions that are not separated into subspaces. Thus there are no specified internal symmetry representations and no space-times within them.[25] There are no particles. There is no dynamics. There is no spatial separation within them—each space can be viewed as a point space. Only their "essence" exists. See chapter 11.

1.3.2 Octonion Octonion Maxiverse Space (4)

Instances of this space, the Maxiverse, cannot be constructed by the annihilation of spaceless space fermions since no fermions exist in the spaceless spaces. See Fig. 1.3. This space has 4096 dimensions, which includes a 10 quaternion octonion space-time. It has urfermions (a type of fermion described later) with 32-spinors. It has internal symmetries and bosons that result from the upward extension of Megaverse space internal symmetries (that extend, in term, from QUeST universe internal symmetries). It has dynamical evolution. It can create Megaverse space (5) instances through urfermion-antiurfermion annihilation. See chapter 3 for more details. Only one instance (particle) of Maxiverse space is created. See chapter 10.

1.3.3 Quaternion Octonion Megaverse Space (5)

Instances of this space (Megaverse space) can be constructed by the annihilation of octonion octonion space fermions. See Fig. 1.3. This space has 1024 dimensions, which includes an 8 complex octonion space-time. A Megaverse instance has a location and momentum in the octonion octonion (4)'s space.-time. It has fermions with 16-spinors. It has internal symmetries that result from the upward extension of QUeST universe internal symmetries. It experiences dynamical evolution. It can create QUeST universe spaces through fermion-antifermion annihilation. See chapter 6 for more details. Any number of instances (particles) of Megaverse space can be created. We know of one Megaverse at most.

1.3.4 Complex Octonion Universe Space (6)

Instances of this space (universe space) can be constructed by the annihilation of quaternion octonion (5) space fermions. See Fig. 1.3. This space has 256 dimensions, which includes a 4 octonions space-time. A universe instance has a location and

---

[24] We explore these spaces from a physical perspective. It is also possible to view these spaces as analogues of Philosophical-Religious phenomena. The three spaceless spaces can be viewed as an analogue of a deity; and the 64 dimension Octonion Octonion space with 10 dimensions and 32-spinor fermions could be viewed as an analogue of the concept of 10 Sefirot with 32 emanations in Judaism. See the *Sefer Yetzirah* – The Book of Creation. Also see Chapter 17.

[25] We use the absence of space-times to end (truncate) the sequence of spaces that are connected by fermion-antifermion annihilation as discussed later.

momentum in a Megaverse's space.-time. It contains fermions with 4-spinors. It has QUeST internal symmetries (consistent with UST) It experiences dynamical evolution. It can create minispaces through fermion-antifermion annihilation. See chapter 2. Any number of instances (particles) of universe space can be created. We know of one universe at present.

1.3.5 Quaternion Point Universe Minispace (7)

Instances of this space can be constructed by the annihilation of complex octonion space fermions in space 6. See Fig. 1.3. This point space has a real-valued 4-dimension space-time. See chapter 12. Any number of instances (particles) of this space can be created.

1.3.6 Quaternion Point Universe Minispace (8 and 9)

Instances of space 8 can be constructed by the annihilation of complex octonion space fermions in space **7**. See Fig. 1.3. This point space has a real-valued 4-dimension space-time. See chapter 12.

Space **9** is constructed by the annihilation of real space-time fermions in space **8**.. See Fig. 1.3. This point space has a real-valued 4-dimension space-time. (chapter 12.) Any number of instances (particles) of this spaces can be created.

1.3.7 Quaternion Point Universe Minispace (10)

Instances of this space can be constructed by the annihilation of fermions in space **9**. See Fig. 1.3. This point space has no space-time, by construction, to preclude an infinite sequence of nested minispaces. A minispace has a location and momentum in a universe's space.-time. It has no fermions or bosons. It can have a subset of QUeST internal symmetries. It does not experience dynamical evolution. See chapter 12. Any number of instances (particles) of this space can be created.

1.3.8 Comments

Instances of spaces 7 – 10 would be created within a universe space instance. An instance of space 7 might be possible in a universe space (6) such as our universe, and might be detected experimentally. Instances of spaces 8 – 10 would appear to be unlikely to be produced or found

The dimension of spaces 7 – 10 total to 64. This number would be the dimension of an octonion coordinate space of octonion coordinates $8 \times 8 = 64$, thus completing the progressive regular pattern of spaces seen in spaces from 0 through 6. The array size 8×8 is not present in Fig. 1.4 because a type 6 universe has fermions with 4-spinors. A space of type 7 can only have instances generated by type 6 fermion-antifermion annihilation. Thus it has a 4×4 dimension array resulting.

If instances of space 7 were derived from space 6 by fermion-antifermion annihilation, then space 7 would have been

7 Octonion (8)	Octonion	8×8	**An invalid alternative**

replacing the 7, 8, 9, and 10 spaces.

The regularity of the pattern of regularly increasing dimension array sizes by factors of 4, and of the progression of coordinate types expressed in terms of octonions, in Fig. 1.4 is very encouraging.

1.4 The Pattern of the Ten Spaces

Spaces 1 through 6 in Fig. 1.4 show an orderly progression of spaces based on the coordinate type and number of coordinates in the second and third columns. Instances of spaces 4 through 10 are linked through fermion-antifermion annihilation as we will see later. The spectrum of spaces is bound above and below by spaces with no space-times and thus no particles or dynamics or fermion-antifermion annihilation.

The three top spaces labeled 1, 2, and 3 are present to round out the coordinates up to "octonion octonion octonion" and to "fill" the block Superverse space of Fig. 1.3.

While spaces 7 and 8 are needed to limit the space specrtrum, spaces 9 and 10 are not required. However we found them useful to create a unified Superverse space consisting of all 10 spaces joined together.

1.4.1 Unification of the 10 Spaces in the Superverse

The 10 spaces are interconnected as shown in Figs. 1.1, 1.2 and 1.3. We will see the interconnections in detail later. If we consider the union of the 10 spaces we find the total number of dimensions totals to 349,504.

The set of 10 spaces can be viewed as subspaces of a complex octonion octonion octonion *Superverse* space with coordinates numbering complex octonion octonion octonion or 1024. Each of the 1024 dimension coordinates is a 1024-vector of the complex octonion octonion octonion type. Fig. 1.6 shows this Superverse space in block-diagonal format upon breakdown to the 10 spaces.

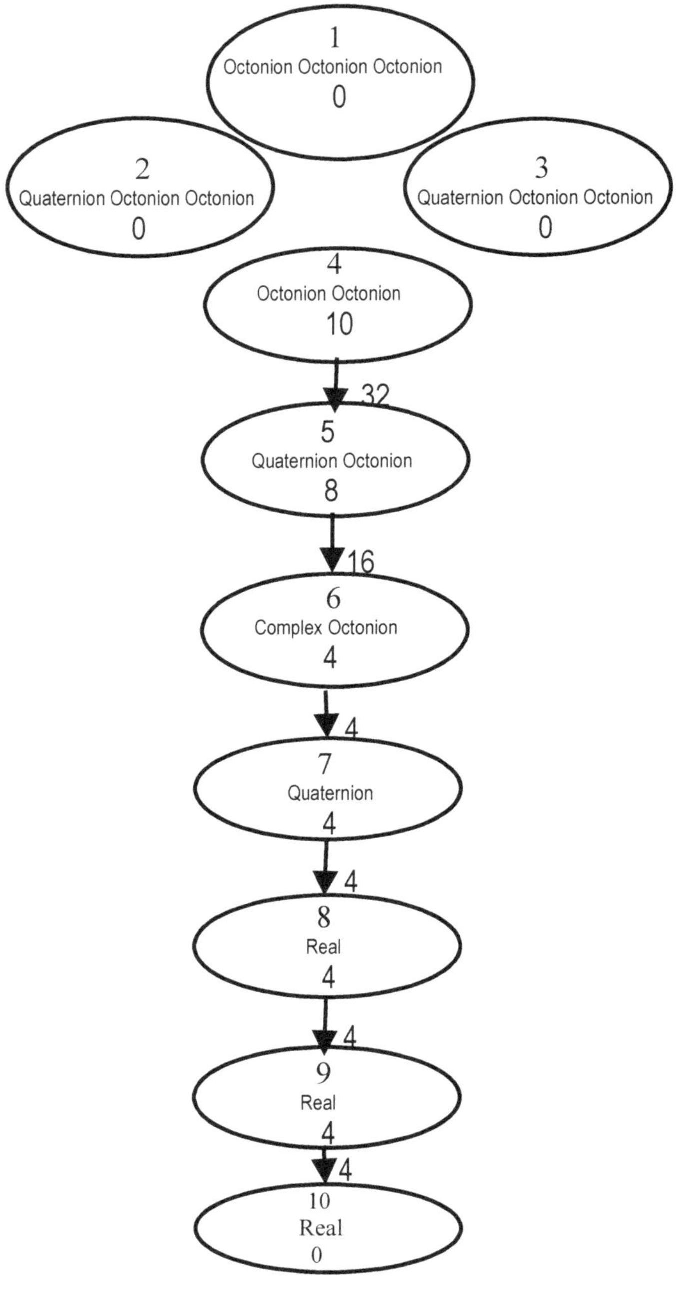

Figure 1.5. The descent of the ten spaces with their spectrum number and their space-time dimensions indicated within each oval. The number of spinor components for each fermion - antifermion pair that annihilates to produce the "next space down" is specified next to each arrow for the seven lower spaces.

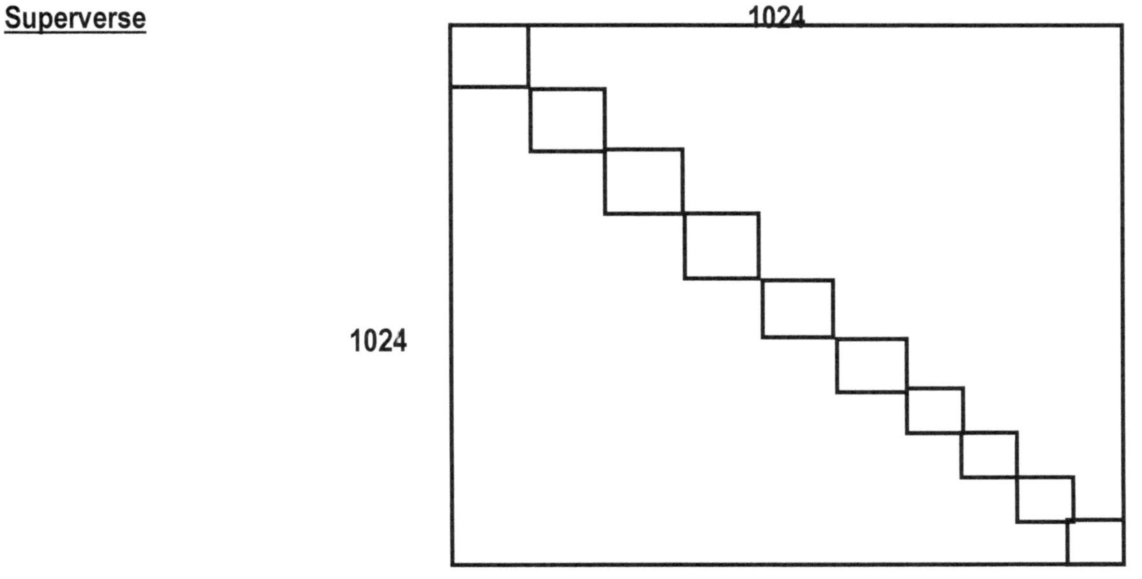

Figure 1.6. The 10 spaces of Fig. 1.4 in block form within a 1024 by 1024 dimension space – the Superverse space. The 10 spaces are 512 by 512, 256 by 256, …, 4 by 4. The shapes of the spaces diagrammed in this figure is not significant.

The 1024 × 1024 Superverse is broken to the 10 spaces in Fig. 1.6. An analogy for the formation of the Cosmos based on the *Sefer Yetzirah* is presented in detail in chapter 17 for the Superverse and its 10 subspaces. It provides a possible justification for Octonion Cosmology.

1.5 Cosmological Evolution

The spectrum of spaces has an evolutionary character. The 3 spaceless spaces are static. They serve to close the spectrum of spaces (an upper limit) while maintaining the mathematical trend of coordinate octonions.

The seven spaces 4 - 10 are needed for the physical Cosmos. The Complex Octonion Octonion (space 3) and the Minispace (space 10) mark the zero space-time dimension endpoints and thus the spectrum limits. These spaces are depicted in Fig. 1.4.

Turning now to the evolutionary sequence we find:

1. The Superverse space (type 0) exists without space-time, particles, or dynamics. It is broken to the tensor product of 10 subspaces.

2. Spaces 1, 2, and 3 are static without space-times, particles, or dynamics.

3. The Octonion Octonion (Maxiverse) Space instance must exist forever. It is populated with elementary particles and interactions. After an unspecified time, an urfermion-antiurfermion annihilation may produce a Megaverse space (type 5) instance. (There may be several/many Megaverse instances spread in time.)

4. A Megaverse space instance (particle) contains elementary particles and interactions. After an unspecified time a fermion-antifermion annihilation may produce a QUeST universe instance of type 6. There may be several/many universes over Megaverse time.

5. A QUeST universe contains the fermions, bosons and interactions of the Unified SuperStandard Theory.

6. Fermion-antifermion annihilation can produce a minispace (types 7, 8, 9) each containing a space-time, elementary particles and interactions..

7. Type 10 space has no space-time. Since it has no space-time it can have no particles or interactions. It is empty. It provides a lower limit to the spectrum of spaces. It also has topological significance.

Minispaces have not been found experimentally. Considering their presumed rarity, and their lack of an experimental signature, the lack of experimental confirmation is understandable. If minispaces were detected they would be significant evidence for Octonion Cosmology! Miniverses of types 7, 8, 9, and 10 could be produced by fermion-antifermion annihilation in principle but are extraordinarily unlikely to be produced.

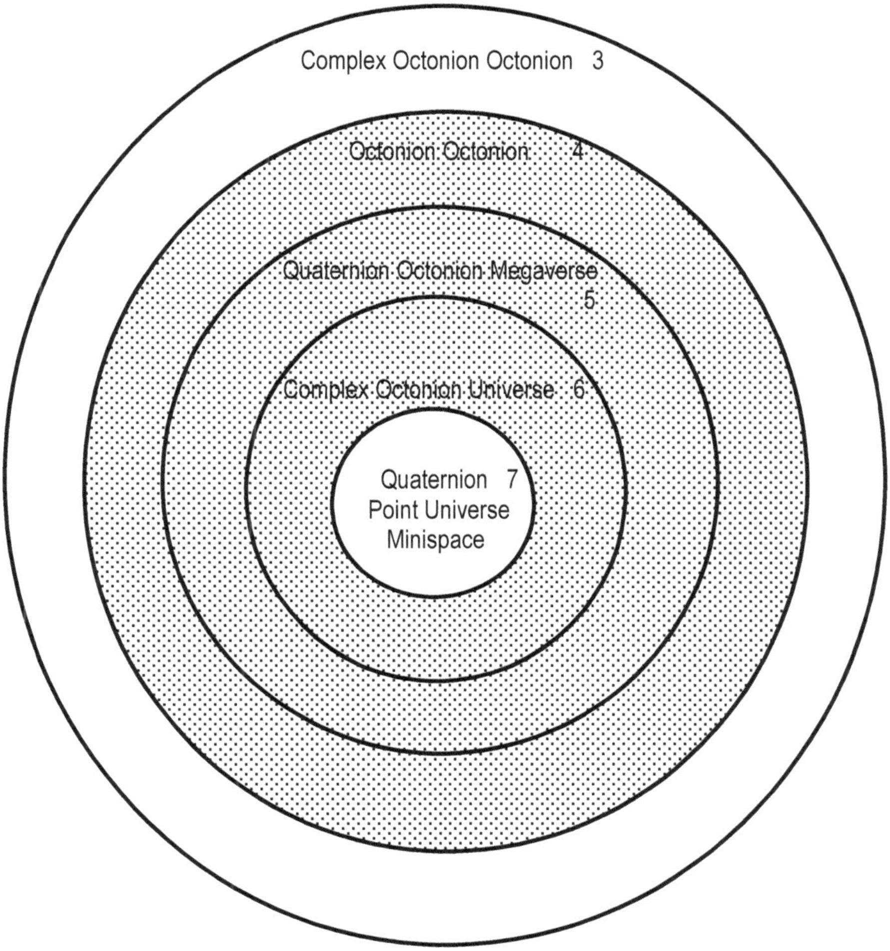

Figure 1.7. The five spaces relevant for the physical Cosmos. The three colored regions have non-zero dimension, and have a non-trivial space containing fermions and bosons with interactions.

2. QUeST for Universes

Earlier this year (2020) the author published a hypercomplex-based theory called QUeST that had the noteworthy feature of directly leading to his Logic-based Unified SuperStandard Theory (UST) of Elementary Particles. QUeST was initially based on a 32 complex quaternion dimension space with 16·16 = 256 dimensions.

Given the success of the octonion spectrum of Chapter 1 (and Blaha (2020k)) we now view QUeST as based on complex octonion coordinates (16) with each coordinate being a complex octonion. Again there are 256 dimensions arranged in a 16 × 16 array. Since we treat hypercomplex numbers solely as a source of dimensions, and exclude use of octonion (and quaternion) units algebra, the above specifications are equivalent.

They each lead to a set of internal symmetries, the UST symmetries, and a four octonion space-time.[26,27] This chapter describes features of QUeST. It consists of material from Blaha (2020c) – (2020j).

2.1 Complex Octonion Space of Complex Octonion Dimensions

Fig. 1.4 lists the space of universes as:

6 Complex Octonion[28] (16)	Complex Octonion	16 × 16	4 octonion

Since Complex Octonions have 16 components this space has 16·16 = 256 dimensions (coordinates). Fig. 2.1 displays the 16 × 16 dimension array. This dimension array is specifies the dimensions of the QUeST formulation.

Fig. 2.1 uses a "dot" • to represent each dimension (a form of Mathematical Picture Language). The dimensions of the space are not assigned physically until they are mapped later to internal symmetry group fundamental representation dimensions and to space-time dimensions. Rather than create a cumbersome coordinate-based notation we choose to use •'s.

The array can also be characterized as a thirty-two complex quaternion space with 32 × 8 = 256 dimensions by moving the 9th through 16th columns in Fig. 2.1 below the sixteen 1st through 8th columns forming a 32 row array. We will alternate between these representations since they are equivalent being based on octonions *without using octonion units algebra*. The 32 × 8 form appears more appropriate when the fermion spectrum is generated in QUeST IF the Dark fermion spectrum parallels the "Normal" fermion spectrum.

---

[26] An octonion time coordinate and three octonion spatial coordinates.

[27] Appendix C contains a larger set of QUeST figures to supplement this chapter..

[28] In our earlier books in 2020 we also designated this 256 dimension space (6) as a 32 complex quaternion space.

Figure 2.1. The 16 × 16 dimension array used by QUeST. Each • represents a dimension.

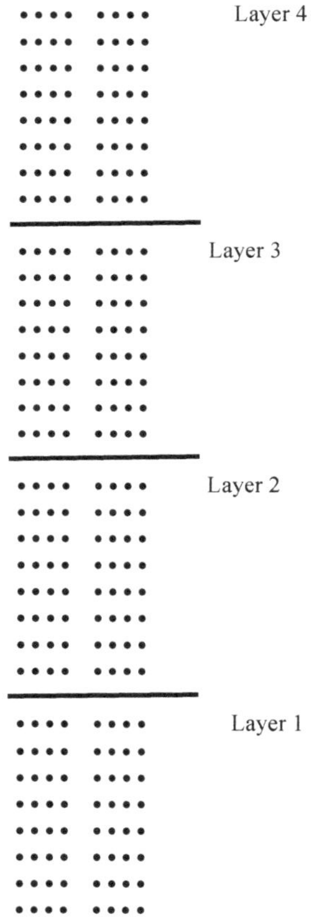

Figure 2.2. The 32 octonions form of the dimension array subdivided into 4 layers of 8 rows..Each layer will be seen to map to a block of fundamental group representations as shown in Figs. 2.3 and 2.4.

In particular, the 32 × 8 form of the array is useful because it brings out the four layers of fermions that appear when the array is subdivided into four layers (8 rows each) of fundamental group representations. The Unified SuperStandard Theory[29] implied by QUeST has a matching four layers of fermions. The shape of the spectrum of fermion masses, the ordering of their masses in particular, would appear to favor a 32 × 8 form since particles of the other three layers have not been found. No interactions are presently known between the Normal and Dark fermions.

2.1.1 Layers

The Unified SuperStandard Theory (UST) of the author divides the fundamental groups and fermions into four layers. The use of layers is due to the appearance of conservation laws of the free field approximation of the UST. It leads to four U(4) symmetry groups that we have called the Layer groups.[30] Each layer has a full set of fundamental groups and a corresponding set of fundamental fermions.

The subdivision of Fig. 2.1 into four layers of 8 rows appears in Fig. 2.2.

2.1.2 Fundamental Group Representations

The dimension array must be partitioned into sets of dimensions for fundamental group representations. Based on the known groups of the Standard Model and the additions in UST we can map dimensions to fundamental group representations.

We use the maps in Figure 2.3 to define the group ↔ dimension map:

U(1)	↔	2 real dimensions
U(4)	↔	8 real dimensions
U(2)	↔	4 real dimensions
SU(3)	↔	6 real dimensions
U(1)⊗SU(2)	↔	4 real dimensions

Figure 2.3.. Map between fundamental group representations and their dimensions. (Dimensions can also be represented as complex.)

The map of the QUeST dimension array to group representations appears in Fig. 2.4. It implements the structure of QUeST and UST that is based on the logic in the case of UST.

2.1.3 Partition of 32 × 8 Dimension Array

It is also based on the footprint of the Megaverse spinors in a fermion-antifermion annihilation that generates a QUeST universe. Chapter 4 describes the annihilation process in detail. See Figs. 4.4 and 4.5 for the form of the footprint generated by the annihilation event.

---

[29] Blaha (2020d) and earlier books.

[30] The Layer groups are discussed in section 5.5 and in detail in Blaha (2020c).

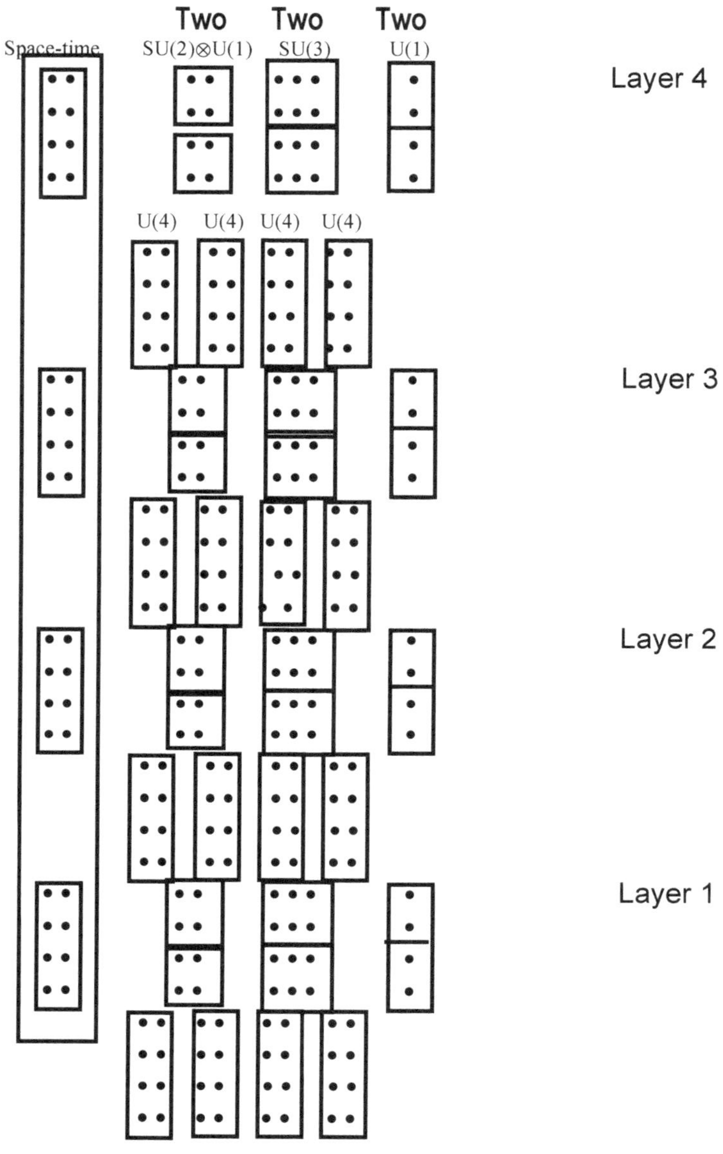

Figure 2.4. The four layers of QUeST internal symmetry groups (and space-time) for the 32 octonion dimension form of space. Note: each row has an 8 • octonion. Note the left column of blocks combine to specify a 4 dimension octonion space-time. Note each layer requires 64 dimensions.. *Note the duplication (using the "Two" label) of each symmetry in each layer.*

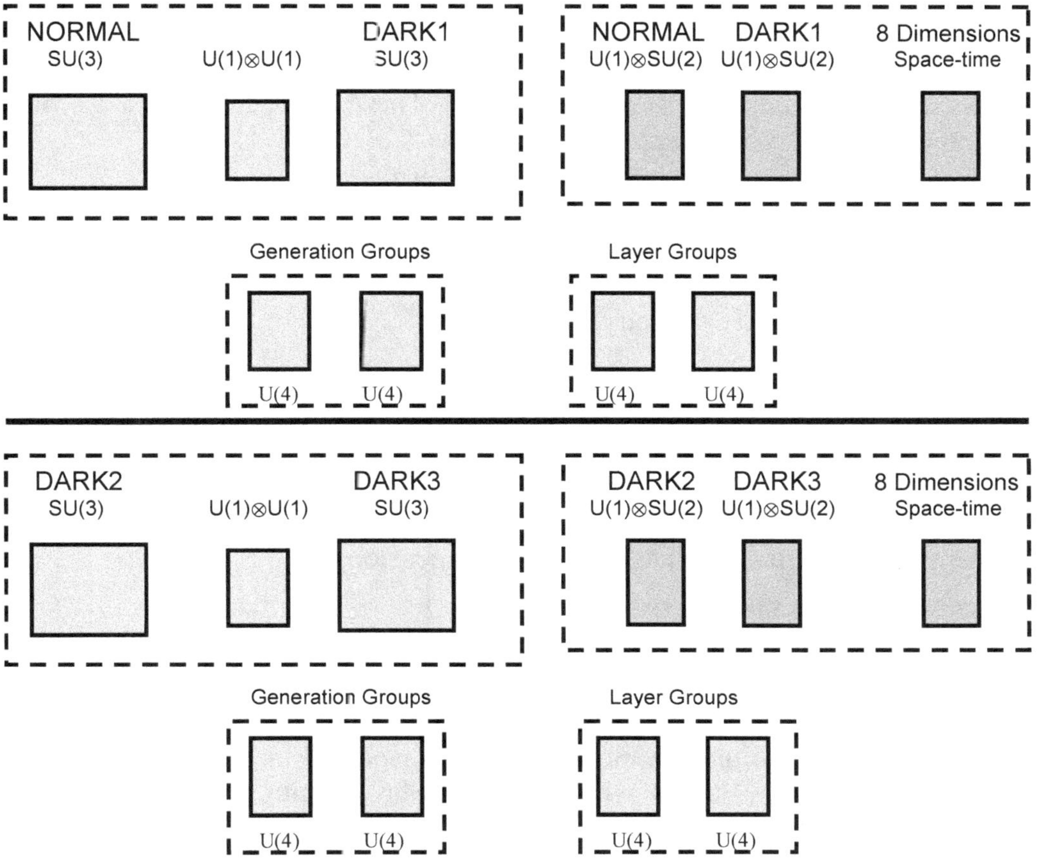

Figure 2.5. Sets of 16 dimension blocks for two of four layers of the 32 × 8 array. Each "dashed" block is a 4 × 4 = 16 array of dimensions. The solid line separates the 64 dimension blocks. This set of 8 blocks contains the 8×16 = 128 dimensions of two of the four layers.

The QUeST dimension array is partitioned by the "footprint" into 4 × 4 = 16 dimension blocks within 8 × 8 = 64 dimension blocks.[31] Fig. 2.5 displays the three types of 16 dimension blocks: A. U(4)⊗U(4), B. U(1)⊗SU(2)⊗U(1)⊗SU(2) and C. SU(3)⊗SU(3)⊗U(1)⊗U(1)⊗(Space-time). These blocks are repeated in the four layers.

QUeST has eight blocks of type A, four blocks of type B and four blocks of type C. UST has the same internal symmetry groups as the sixteen blocks of QUeST.

---

[31] The 8 × 8 = 64 dimension blocks arise as the footprint of the urfermion-antiurfermion annihilation in the Maxiverse that generates an UTMOST Megaverse instance. See section 8.3.

2.1.4 Map Between Octonion Cosmology and UST

The partition of the QUeST dimension array into 16 dimension blocks raises the question of the origin of the three types of blocks shown in Fig. 2.5. In section 5.2 below we use a correspondence principle that relates the structure of the blocks to UST. The blocks initially could be U(8) representations. These representations are broken to the three types in Fig. 2.5 by comparing to UST.

The QUeST dimension structure (Fig. 2.4) results.

2.3 QUeST Symmetries

2.3.1 QUeST Internal Symmetry Groups

The QUeST internal symmetry group is

$$[SU(2)\otimes U(1)\otimes SU(3)]^8\otimes U(4)^{16}\otimes U(1)^8 \qquad (2.1)$$

by Fig. 2.4 above.

2..3.2 QUeST Space-Time

The space-time of QUeST is 4 octonion dimensions.

2.3.3 QUeST/UST Fermion Spectrum

Given the form of the internal symmetries in QUeST we can determine the fermions in the fundamental group representations as shown in Fig. 2.6. The set of Internal Symmetries in Fig. 2.4 determines the set of fermions in fundamental representations. The resulting fermion spectrum is displayed in Fig. 2.6. The fermions are tentatively ordered by layers with the lowest layer being known in part. Each Normal layer part is tentatively paired with a (possibly higher mass) Dark part.

2.3.4 Generation and Layer groups of UST, QUeST and BQUeST

The Generation groups mix the fermion generations of normal and Dark sectors of each layer. The lines on the left side of Fig. 2.6 display Generation group mixing within each layer.

Layer groups mix fermions in all four layers for each of the four generations individually. (See right side of Fig. 2.6.) There are eight Layer groups: two Layer groups for Normal and Dark sectors for each generation.

See Blaha (2020c) for a detailed discussion.

2.3.5 Fermion- Dimension Duality

Fig. 2.7 shows a 1:1 relation between QUeST dimensions and the fundamental fermions of QUeST and UST. The fundamental fermions are those fermions in the fundamental representation of the respective groups. This duality is the basis of the BQUeST and BMOST one-dimension theories implying QUeST and UTMOST respectively.

The Fermion Periodic Table

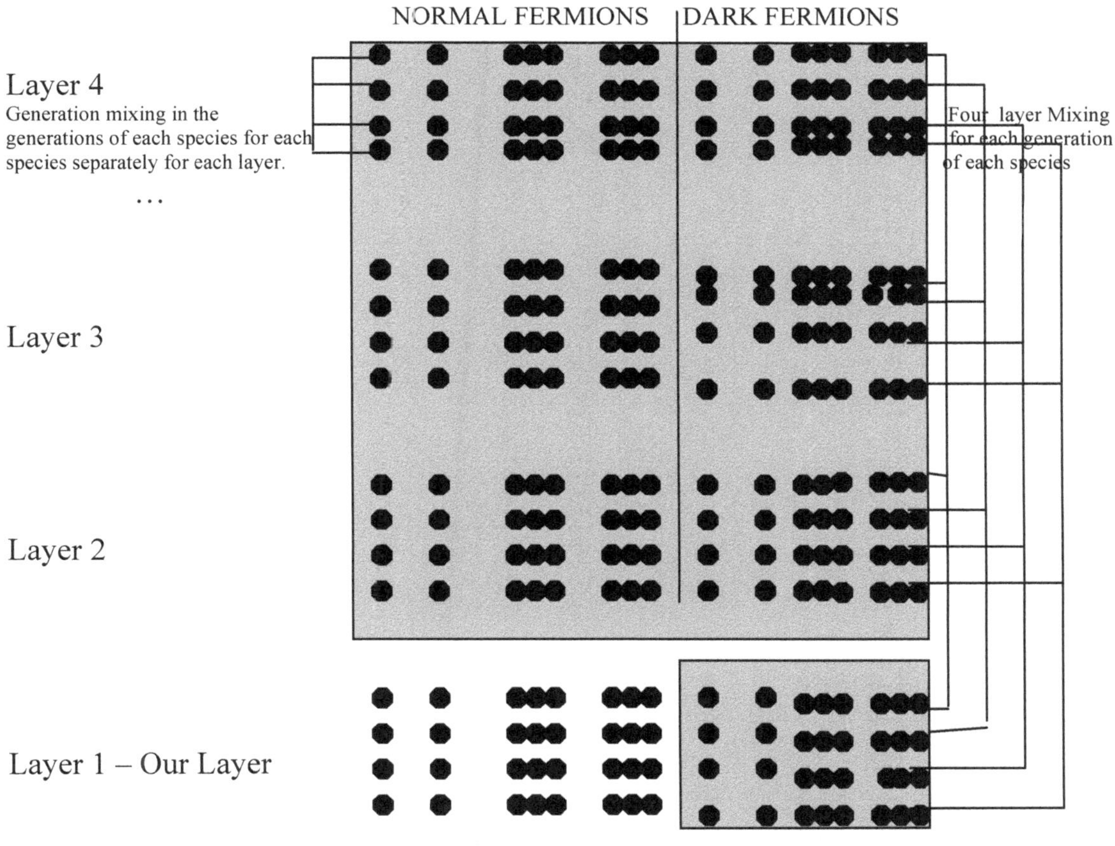

Figure 2.6. Fermion particle spectrum and partial examples of the pattern of mass mixing of the Generation group and of the Layer group. Unshaded parts are the known fermions with an additional, as yet not found, 4th generation. The lines on the left side (only shown for one layer) display the Generation mixing within each layer. The Generation mixing occurs within each layer using a separate Generation group for each layer. The lines on the right side show Layer group mixing (for Dark matter) with the mixing among all four layers for each of the four generations individually. There are four Layer groups for Normal matter and four Layer groups for Dark matter.. There are 256 fundamental fermions. QUeST and UST have the same fermion spectrum.

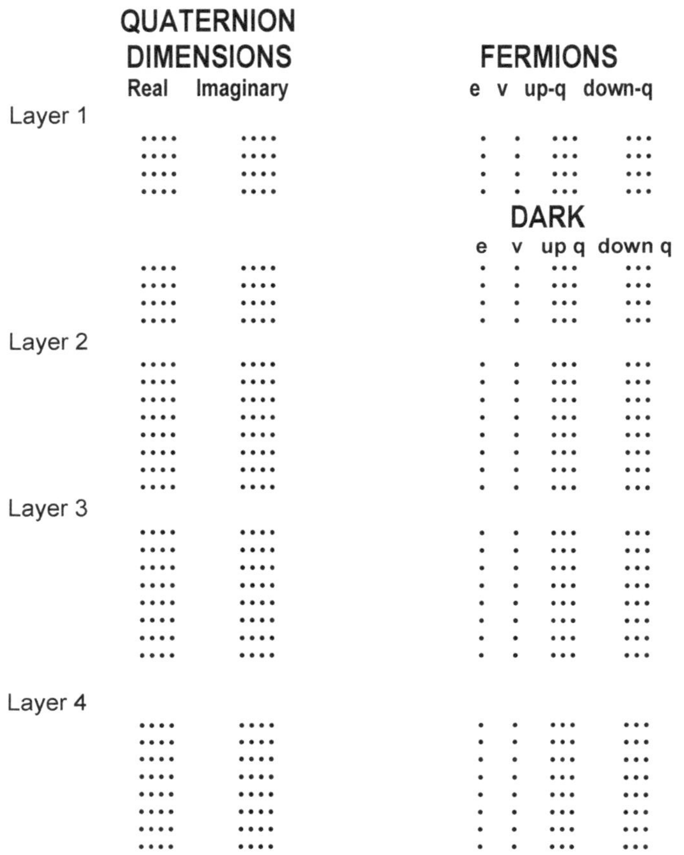

Figure 2.7. Fundamental fermions have a 1:1 correspondence with QUeST dimensions. Note the number of dimensions in each row is 8 – the number of dimensions in an octonion. Correspondingly the number of fermions in each row is 8 – a suggestive similarity. Each layer has four normal fermion generations and four Dark fermion generations. Each dot (pebble) represents a dimension in the left part of the figure and a fermion in the right part.

2.4 16 Dimension Block Pattern of Fermions

The possible U(8) 4 × 4 = 16 block structure described in section 2.1.3 has an analogous pattern of fundamental fermions. It is shown in Fig. 2.8. Each block holds 16 fermions in four rows. Each block has a lepton and 3 up or down quarks in 4 generations.[32] The blocks are structured in a manner suggested by ElectroWeak theory

The form of the blocks suggests a new form of symmetry breaking arrays of quarks and leptons.

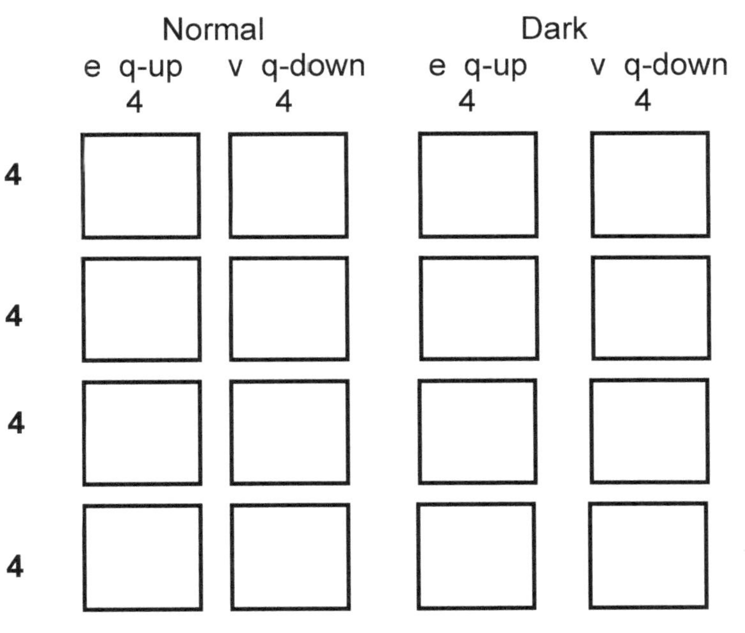

Figure 2.8. A block form of a 16 × 16 QUeST fermion array with each block row corresponding to one layer. Each block contains four generations of fermions. The result is 4 × 4 blocks. The label e q-up indicates a charged lepton – up-type quark pair, v q-down indicates a neutral lepton – down-type quark pair, and so on. Note the blocks can be reaaranged into a 32 × 8 form without physical consequences at this level of discussion since the right two columns and the lowest two rows are all Dark at present.

See Appendix C for more details of the QUeST formulation. Chapter 5 discusses the QUeST – UST joint formulation and shows the remarkable match of QUeST and UST internal symmetry groups and particle spectrums.

[32] The lepton-3 quark pattern is suggestive of a Lorentz 4-vector representation with the lepton corresponding to a time coordinate and the three quarks corresponding to spatial coordinates.

3. The Derivation of Quaternion QUeST Universes

QUeST, and the Unified SuperStandard Theory (UST) that it implies, is based on a complex octonion space with 256 dimensions, which appears in Fig. 1.4 as

6 Complex Octonion[33] (16)	Complex Octonion	16×16	4 octonion

The questions arise: Is there a deeper basis for QUeST with much fewer dimensions? Can QUeST, which describes a universe, be a subspace that originates as a "particle" in a larger space that we call the Megaverse (Multiverse)? Questions of this sort have led the author to find a deeper theory that implies 256 dimension QUeST. The theory, called BQUeST, is described in Blaha (2020i) and (2020j).

This chapter describes BQUeST in much greater detail. The process for the birth of a universe instance begins in a fermion-antifermion annihilation in the 8-dimension Megaverse. A Megaverse fermion, when off shell, has 16 independent spinor components. The annihilation leads to the $16 \cdot 16 = 256$ internal dimensions of a QUeST universe.

3.1 BQUeST Origin

We begin with the assumption that a QUeST universe originates in an UTMOST Megaverse.[34] Since the UTMOST Megaverse has 8 space-time dimensions, an eight dimension "seed" fermion has 16 spinor components. (There are $256 = 16 \times 16$ Dirac matrices, denoted G_{ab} for a, b = 1, 16, and eight 16×16 Dirac γ matrices.)

We view a Megaverse instance as the origin of a QUeST universe instance. We picture an off shell seed fermion and its off shell seed antifermion as appearing as fluctuations in the Megaverse vacuum. They "annihilate" to create a QUeST universe particle.[35] The universe particle, being in a Big Bang state of enormous energy, then proceeds to expand.

We will show the 16 independent spinor components of the 8 dimension off shell seed fermion and of the off shell anti-seed fermion generate the 16×16 dimension array of QUeST.[36]

---

[33] In our earlier books in 2020 we also designated this 256 dimension space (6) as a 32 complex quaternion space.

[34] QUeST universes are described in Appendix C. The UTMOST Megaverse is described in Appendix E.

[35] Appendix D provides experimental and theoretical evidence for a particle view of universes. Comments by DeWitt and others on quantum universes support this view.

[36] The UTMOST Megaverse has an eight complex octonion space-time. We choose to extract the real coordinate of each complex octonion 16-tuple to construct a real 8 dimension space-time within which seed fermion dynamics occurs. This is analogous to the case of complex 4 dimension space-time being restricted to real space-time in conventional quantum field theory.

3.2 A Model Dynamics of the Seed – Anti-Seed Generation of a QUeST Universe

The seed fermion must have a dynamics that enables it to generate the 256 dimension array. In this section we describe a model dynamics for the annihilation of a seed–antiseed state into a scalar universe instance (particle). The model must have certain features that accomplish the goal:

1. Since the target dimension array is not symmetric the PseudoQuantum formulation[37] of Quantum Field Theory will be seen to be required.

2. Since the quantum field theory is in 8 dimensions we must use Two-Tier coordinates,[38] which eliminate all divergences in perturbation theory using exponential damping of all integrations.

These extensions of quantum field theory were developed by the author in the 1970's and the early 2000's. They are both described in Appendix A together with references to earlier papers and work.

We define a model lagrangian for the seed fermion sector. The seed fermion is represented by *two* quantum fields[39] ψ_1 and ψ_2. The scalar[40] universe particle has two quantum fields A_{1ab} and A_{2ab} with indices a and b.

The lagrangian terms are

$$\mathscr{L} = F^{1\mu\nu} F^2{}_{\mu\nu} + \overline{\psi}_2 (i\gamma\cdot\partial - m)\psi_1 + \overline{\psi}_1 (i\gamma\cdot\partial - m)\psi_2 - e_0\overline{\psi}_2 A_2\psi_1 - \overline{\psi}_1 A_1\psi_2 \qquad (3.1)$$

where m is the mass of the seed fermion, and where $A_i = A_{iab}G_{ab}$ summed over a and b.

To insure finite perturbation series results (in 8 dimensions) the fields are all functions of Two-Tier coordinates X having a vector quantum field $Y^\mu(y)$:

$$X^\mu(y) = y^\mu + i\, Y^\mu(y)/M_c^2 \qquad (3.2)$$

where M_c is an extremely large mass.

3.2.1 U(16) Species Group Symmetry for Dirac Matrix Indices

The lagrangian terms (plus additional universe terms) are invariant under a U(16) Species symmetry group. Appendix B describes the analogous U(4) Species[41] symmetry for four dimension fermions that we found for UST.

---

[37] S. Blaha, Il Nuovo Cimento **49A**, 35 (1979) and Appendix A.

[38] See Appendix A.

[39] We use a PseudoQuantum scalar pair of fields also. A scalar particle is a model of the universe. See eqs. 61 – 90 in Appendix 1-C of Blaha (2020j). The Two-Tier formulation eliminates all perturbation theory divergences in four dimensions including the notorious fermion triangle divergence. It also eliminates all perturbation theory divergences in eight dimensions.

[40] Current experimental evidence suggests the universe is not rotating.

[41] The U(4) Species symmetry group is part of UST. It originates in the Complex General Reloativity of four dimension coordinates as shown in Appendix A and Blaha (2020c) as well as earlier books.

Appendix B shows the form of the trial lagrangian terms for establishing the Species group of UST in conventional quantum field theory:

$$\bar{\psi}'(x)[i\gamma_\mu'(x)(\partial/\partial x_\mu - ig_8 A'_{Sk}{}^{\mu}(x)G_{Sk}) - m]\psi'(x) \tag{51.2}$$

where the matrices G_{Sk} are the 16 Dirac matrices of 4 dimension space, where k ranges from 1 through 16, and where the $A'_{Sk}{}^{\mu}$ fields.are 16 gauge fields for the Species group.

In developing the U(4) Species group symmetry for four dimension UST we related the indices of the Dirac matrices $[G_{Sk}]_{ab}$ where a, b = 1, 2, 3, 4 to the space-time coordinates using vierbein transformations. In this regard Appendix B states:

A complex transformation of types II and III in Fig. 50.1 has the form:

$$U(x'')^{\mu}{}_{\nu} = w^{\mu}{}_{a}(x'')[\exp(i\textstyle\sum_k \Phi_k(x'')\tau_k)]^{a}{}_{b}\,1^{b}{}_{\nu}(x'')$$
$$U^{-1}(x'')^{\mu}{}_{\nu} = w^{\mu}{}_{a}(x'')[\exp(-i\textstyle\sum_k \Phi_k(x'')\tau_k)]^{a}{}_{b}\,1^{b}{}_{\nu}(x'')$$

where τ_k is a U(4) generator matrix. Its infinitesimal transformation is approximately

$$U(x'')^{\nu}{}_{\beta} \approx \delta^{\nu}{}_{\beta} + i\textstyle\sum_k \Phi_k(x'')[\tau_k]^{\nu}{}_{\beta} \tag{50.12}$$
$$U^{-1}(x'')^{\nu}{}_{\beta} \approx \delta^{\nu}{}_{\beta} - i\textstyle\sum_k \Phi_k(x'')[\tau_k]^{\nu}{}_{\beta}$$

using the *vierbein* flat space-time limits

$$w^{\mu}{}_{a}(x'') \approx \delta^{\mu}{}_{a} \tag{50.12a}$$
$$1^{b}{}_{\nu}(x'') \approx \delta^{b}{}_{\nu}$$

where

$$\Phi_k(x) = \int^{x} dy_\lambda\, A_{Rk}{}^{\lambda}(y) \tag{50.13}$$

Then

$$\Gamma_R{}^{\sigma}{}_{\lambda\mu} = -\tfrac{1}{2}i\{\textstyle\sum_k A_{Rk}(x'')_{\mu}[\tau_k]^{\sigma}{}_{\lambda} + \sum_k A_{Rk}(x'')_{\lambda}[\tau_k]^{\sigma}{}_{\mu}\} \tag{50.14}$$
$$= A_R{}^{\sigma}{}_{\mu\lambda} + A_R{}^{\sigma}{}_{\lambda\mu}$$

(summed over k) with the matrix $A_R{}^{\sigma}{}_{\mu\lambda}$ given by

$$A_R{}^{\sigma}{}_{\mu\lambda} = -\tfrac{1}{2}i\textstyle\sum_k A_{Rk\mu}[\tau_k]^{\sigma}{}_{\lambda} \tag{50.15}$$

with $A_R{}^{\sigma}{}_{\mu\lambda}$ transformable to matrix row and column numbers

$$A_{Rflat}{}^{\mu a}{}_{b} = A_{Rflatk}{}^{\mu}[\tau_k]^{\sigma}{}_{\lambda}\delta_{\sigma}{}^{a}\delta^{\lambda}{}_{b} \tag{3.3}$$

using the flat space-time vierbein values, and so $A_{Rflat}{}^{a}{}_{\mu b}$ may be written in matrix form as

$$A_{R_{flat\mu}} = -\tfrac{1}{2} i \sum_k A_{R_{flatk\mu}} \tau_k \qquad (50.16)$$

In the flat space-time limit the $A_{Rk}{}^{\lambda}(y)$ becomes the Coordinate Species group U(4) gauge fields $A_{R_{flatk}}{}^{\lambda}(y)$.

The above extract shows that space-time dimensions can be directly related to matrix indices. We will use this feature to map the indices of the universe fields A_{iab} to internal dimensions of the universe particles. These dimensions, which number 256 in the 8 dimension Megaverse space-time, become the 256 dimensions in QUeST. (It should be noted that the UTMOST Megaverse has 1024 dimensions —much more than the 256 dimensions of QUeST.

3.3 The Interaction Generating a Universe Particle

The diagram for the creation process is in Fig. 3.1. A seed fermion and an antiseed fermion annihilate to create a scalar universe particle.[42]

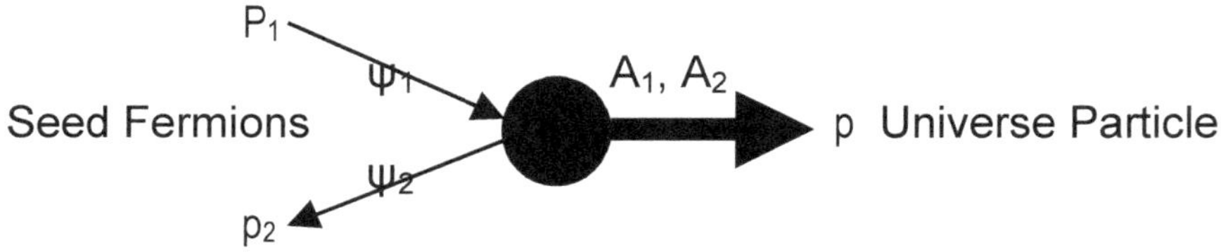

Figure 3.1. Diagram for the transition of seed - antiseed fermions to a universe particle.

The lowest order S-matrix element for this interaction has the form:[43]

$$S = \int d^8x\, [\overline{\psi}_2 A_2 \psi_1 + \overline{\psi}_1 A_1 \psi_2] \qquad (3.4)$$

After evaluating the integration we find S has terms of the form

$$\overline{v}_{2a}(p_2)\, A_{1ab}(p)\, u_{1b}(p_1) \qquad (3.5)$$

Rearranging the expression we obtain the trace expression

$$\mathrm{Tr}\ u_{1b}(p_1)\overline{v}_{2a}(p_2)\, A_{1ab}(p) = U_{ba} A_{1ab}(p) \qquad (3.6)$$

---

[42] The possibility of seed-antiseed annihilation into two universe particles is not excluded This possibility might account for the handedness and preponderance of matter (as opposed to antimatter) in our universe.

[43] Although a Megaverse has 8 complex octonion space-time dimensions we can reduce it to 8 real-valued dimensions.

where U_{ba} is an array composed of u and $\overline{v}$.

Note the interaction can only take place for far off shell seed and antiseed fermions. Then the 16 spinor components of each fermion are independent and can be treated as independent.

Universe production (Fig. 3.1) has momentum conservation due to the off shell fermions. The diagram has two interesting cases:

1. If A_{1ab} and A_{2ab} are both proportional to a Kronecker δ_{ab} then the S matrix expression of eq. 3.4 simplifies to the scalar A_i case.

2. If A_{1ab} and A_{2ab} are not proportional to a Kronecker δ_{ab} then the S matrix expression can be expressed as a sum of terms for each value of a and b:

$$S = \Sigma\, S_{ab} \tag{3.7}$$

where (with no sum over a or b)

$$S_{ab} = \int d^8x\, [\overline{\psi}_{2a} A_{2ab} \psi_{1b} + \overline{\psi}_{1a} A_{1ab} \psi_{2b}] \tag{3.8}$$

The second expression eq. 3.8 suggests each ab pair of values marks a channel from the fermions to the universe particle. Thus we can view the universe particle as receiving 256 “streams” from the fermion-antifermion pair.

In section 3.4 below we suggest the streams “amalgamate” to form a 256 dimension universe.

3.4 U(16) Symmetry of A_i

The lagrangian in eq. 3.1 has a U(16) symmetry that is analogous to the 4 dimension U(4) Species group described in section 51.1 of Appendix B. This U(16) symmetry exists in spinor “space.” The matrix, $[U_{ba}]$ with its 16-spinor indices a and b, has a symmetry under the U(16) transformation:

$$U[U_{ba}]U^{-1} \tag{3.9}$$

U(16) symmetry requires the A_i factors in eqs. 3.5 and 3.6 to also have a U(16) transformation of the form:

$$A_i' = UA_iU^{-1} \tag{3.9a}$$

This symmetry will be seen to impact on the form of the QUeST dimension array.in chapter 2.

3.5 Indices of [A_{1ab}] Transformed to Internal Dimensions

Eq. 3.5 shows the indices of the array, [A_{1ab}], have 16 × 16 = 256 values due to the 16×16 independent values of the products: $u_{1b}(p_1)\overline{v}_{2a}(p_2)$. Eq. 3.3 above shows that indices can be transformed to coordinates.

Therefore we can take the array of indices [ab] and transform them to coordinates using a Megaverse vierbein.[44] We can take the coordinates that emerge as each corresponding to a one dimension line (a stream?) within the universe particle.

Subsequently, or simultaneously, after creation, the energy-momentum within the streams can be distributed/flow into the 4 dimension space-time within the QUeST universe. The internal energy-momentum must sum to zero for momentum conservation. So there will be patches of positive and negative energies. (Negative energies may be absorbed by a vacuum rearrangement.)

The independence of the spinor components of each of the two seed fermion fields guarantees the elements of the array of indices form an array of independent dimensions.

3.6 The Difference between Indices and Internal Dimensions

The array [A_{1ab}] is an array with indices. We now consider the relation of indices and internal dimensions of an entity. First we consider the case of a cube, which can be viewed as an entity with three indices. Its external indices can be taken to label the length, width and height of the cube. Within the cube we can establish a coordinate system with three dimensions labeling points within it. Thus external indices can be viewed as mapping to internal dimensions within.

Perhaps a better example is a Black Hole. It exists in 4-dimension space. Within its event horizon the coordinates of the external space can be continued. It is also possible to replace them with an internal coordinate system of four dimensions. This possibility could be supported by the role of radial dimension externally becoming effectively a time coordinate internally.

We thus conclude the indices of A_{iab} can be viewed as specifying coordinates (*dimensions)* within A_i. We have mapped the seed fermion to a 16 × 16 = 256 dimension array. This array serves as the dimensions of the QUeST space.

3.7 Origin of Universe

We can then envision the possibility that our universe started as the annihilation of a seed fermion and antifermion in the Megaverse. It "acquired" the 256 dimensions of QUeST to form a space that includes our 3+1 space-time at the Big Bang point as well as internal symmetries. This process could proceed, as it likely does, instantaneously. The seed fermion-antifermion pair are necessarily in an 8-dimension UTMOST space-time as we showed in previous books. Thus QUeST universes are directly linked to the UTMOST Megaverse.

---

[44] The Megaverse undoubtedly has an 8-dimension General Relativity. For small neighborhoods of a point the vierbein becomes the identity as eq. 50.12a above shows.

Appendix D shows that the post-Big Bang stage of the universe is analogous to the vacuum polarization of a charged particle, substantiating the interpretation of the universe as a type of particle.

3.8 Nature of the Seed Fermion and Antifermion

The type of the seed fermion and antifermion are one of the possible 1024 fermions in the UTMOST Megaverse. In Blaha (2020e) we showed that QUeST and MOST fermions occur in four species: charged leptons, neutral leptons, up-type quarks, and down-type quarks. Examples are e, ν_e, u, and d. The same set of four species appears in UTMOST as shown in section 3.9 below.

There are four layers of fermions in UTMOST. Each layers has eight sets of four generations of four species. In total there are 128 · 8 = 1024 fermions. A reasonable possibility is the seed fermion is an *electron-type fermion* with the lowest mass of all 1024 fermions based on the masses of the known fermions.

3.9 Demonstration That UTMOST Fermions Occur in Four Species

This section follows Blaha (2020e) applied to UTMOST.

In the Megaverse, the fermion species in UTMOST number four despite the 8-dimension complex octonion nature of UTMOST space-time.

There are four types of boosts in this 7+1 complex octonion-valued space-time that boost a particle rest state to a state of motion with a real energy, and a real-valued or complex octonion valued momentum (spatial coordinates):

1. A boost from rest to a frame with real-valued energy and momentum with $p^{02} - ||\mathbf{p}||^2 > 0$. A "normal charged lepton-like" fermion.

2. A boost from rest to a frame with real-valued energy and momentum with $p^{02} - ||\mathbf{p}||^2 < 0$. A "tachyonic neutral lepton-like" fermion.

3. A boost from rest to a frame with real-valued energy and a complex quaternion valued spatial 7-momentum with $p^{02} - ||\mathbf{p}||^2 > 0$. An "up-type quark-like" fermion.

4. A boost from rest to a frame with real-valued energy and a complex quaternion valued spatial 7-momentum with $p^{02} - ||\mathbf{p}||^2 < 0$. A "tachyonic down-type quark-like" fermion.

where p^0 is the real part of the octonion energy, and where $\mathbf{p}$ is a 7-vector and each of its momentum components $\mathbf{p}_k$ are spatial 16-vector (complex octonion):

$$
\begin{aligned}
\mathbf{p} &= \mathbf{p}_1 + \mathbf{p}_2 + \mathbf{p}_3 + \ldots + \mathbf{p}_7 \\
\mathbf{p}_k &= \mathbf{p}_{kr} + i\mathbf{p}_{ki} + j\mathbf{p}_{k3} + k\mathbf{p}_{k4} + q\mathbf{p}_{k5} + r\mathbf{p}_{k6} + s\mathbf{p}_{k7} + \ldots + t\mathbf{p}_{k16} \qquad (3.10) \\
||\mathbf{p}|| &= \mathrm{sqrt}(\mathbf{p}_r\cdot\mathbf{p}_r + \mathbf{p}_i\cdot\mathbf{p}_i + \mathbf{p}_3\cdot\mathbf{p}_3 + \ldots + \mathbf{p}_{16}\cdot\mathbf{p}_{16})
\end{aligned}
$$

where i, j, k, q, r, s, …, and t are fundamental octonion units, and where $||\mathbf{p}||$ is the norm of $\mathbf{p}$. The $\mathbf{p_k} \cdot \mathbf{p_k}$ terms are 16-vector inner products.

4. Connection of Seed Fermion Structure to QUeST Dimensions Structure

Decisive experimental evidence for the origin of a universe in seed fermion-antifermion annihilation will not be found for many millenniums, if at all. There is a possibility that theoretical support for our proposed origin of universes may arise. The possibility is based on the possibility that the structure of the QUeST dimension array has a "footprint" of the form of the seed fermion-antifermion prior to annihilation.

The hope: The seed fermion-antifermion annihilation that produces a QUeST universe has features that appear to impress themselves on universe structure. This chapter describes this possibility.

4.1 Structure of 16-Spinor u and v

The sixteen 16-spinors of an 8 dimension fermion have the general form:

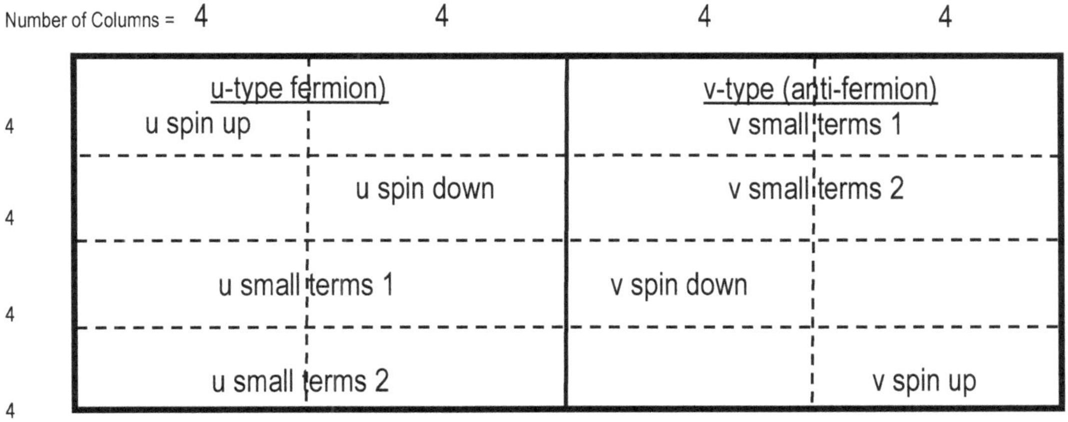

Figure 4.1. The 16 spinors of a 8 dimension spinor. Each spinor has 16 rows.

4.2 Compression of u and v into Composite Spinors

We can combine the eight u-type spinors into a composite spinor to exhibit their structure. Similarly, we can combine the eight v-type spinors into a composite spinor. Fig. 4.2 shows the forms of the composite u-type and v-type spinors.

	u-type Composite		v-type (anti-fermion)	
4	u spin up	u-up	v small terms 1	v-v1
4	u spin down	u-down	v small terms 2	v-v2
4	u small terms 1	u-v1	v spin down	v-down
4	u small terms 2	u-v2	v spin up	v-up

Figure 4.2. Composite u-type and v-type spinors illustrating the structure of each type of spinor based on combining the two columns of each spinor type into one column. Each spinor is subdivided into 4 row sections. The third and fifth columns are symbols for each sector.

The outer product of spinor factors in eq. 1.6 is $u\overline{v}$. Fig. 4.3 contains the form implied by the composite spinors.

u-up–v-v1	u-up–v-v2	u-up–v-down	u-up–v-up
u-up–v-v1	u-up–v-v2	u-up–v-down	u-up–v-up
u-up–v-v1	u-up–v-v2	u-up–v-down	u-up–v-up
u-up–v-v1	u-up–v-v2	u-up–v-down	u-up–v-up
u-down–v-v1	u-down–v-v2	u-down–v-down	u-down–v-up
u-down–v-v1	u-down–v-v2	u-down–v-down	u-down–v-up
u-down–v-v1	u-down–v-v2	u-down–v-down	u-down–v-up
u-down–v-v1	u-down–v-v2	u-down–v-down	u-down–v-up
u-v1–v-v1	u-v1–v-v2	u-v1–v-down	u-v1–v-up
u-v1–v-v1	u-v1–v-v2	u-v1–v-down	u-v1–v-up
u-v1–v-v1	u-v1–v-v2	u-v1–v-down	u-v1–v-up
u-v1–v-v1	u-v1–v-v2	u-v1–v-down	u-v1–v-up
u-v2–v-v1	u-v2–v-v2	u-v2–v-down	u-v2–v-up
u-v2–v-v1	u-v2–v-v2	u-v2–v-down	u-v2–v-up
u-v2–v-v1	u-v2–v-v2	u-v2–v-down	u-v2–v-up
u-v2–v-v1	u-v2–v-v2	u-v2–v-down	u-v2–v-up

Figure 4.3. Outer product array $[U_{ba}]$ (eq. 1.9) of the composite u-type and v-type spinors illustrating the structure of the outer product array of uv's.

Since the annihilation process for the *unpolarized* seed fermion-antifermion pair requires a sum over spin, the four columns of Fig. 4.3 are 16 columns in reality due to the spin summation An 8 dimension fermion has four up-spin values and four down-spin values.

4.3 Block Form of $[U_{ba}]$ Array

We now turn to partiticning the $[U_{ba}]$ and identifying the blocks of the QUeST array implied by the partition. Each column in Fig. 4.3 becomes four columns in $[U_{ba}]$ extended by the summation over the spin of the unpolarized seed fermions. The pattern of array entries is displayed in Fig. 4.4.

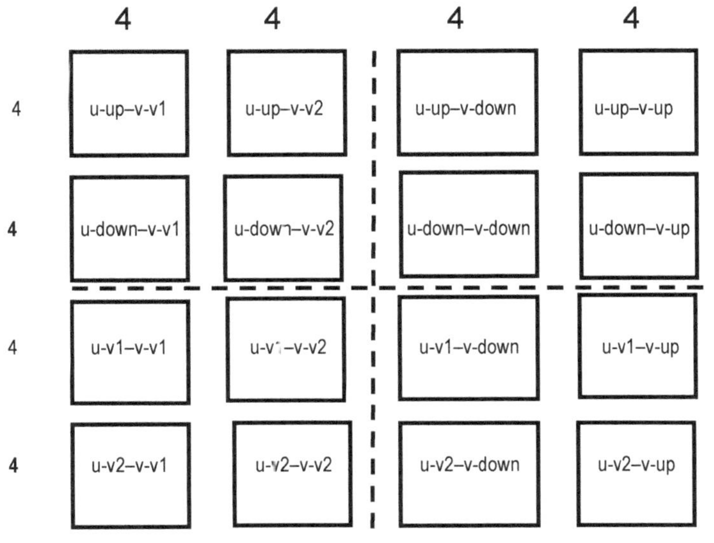

Figure 4.4. Block form of the 16 × 16 $[U_{ba}]$ array. This is also the form of the QUeST dimension array of 256 dimensions. The blocks are divided by dashed lines that separate 64 dimension sections. These sections map to layers in QUeST (UST) as shown in Figs. C.2 and C.3.

The 4 × 4 blocks in the array match the 4 × 4 blocks that we found in earlier books such as Blaha (2020j). Figs. C.9, C.!0, and C.12 show the 4 × 4 block structure of QUeST dimensions and fermions. The sets of four blocks are the four layers of a QUeST and UST. Four layers are evident.

We can rearrange the blocks of Fig. 4.4 to a four layer format (Fig. 4.5) similar to QUeST. Fig. 4.5 shows the block layout of Fig. 4.4 reshaped by moving the right blocks below the left blocks.

Thus we find the structure of the QUeST dimension array (Figs. C.3 and C.4) is a reflection of the structure of the seed fermion-antifermion spinor array. It has 16 sixteen dimension blocks. Each of the above two layers has 128 dimensions within it.

The blocks can be reassembled in pairs to give the form of each of the four layers (Fig. 4.5). Transformed into the four layer equivalent, each of the four layers has 64 dimensions in it. The QUeST dimension array maps directly to our UST theory as shown in chapter 3.

BQUeST yields the QUeST dimension array structure as described in our earlier books!

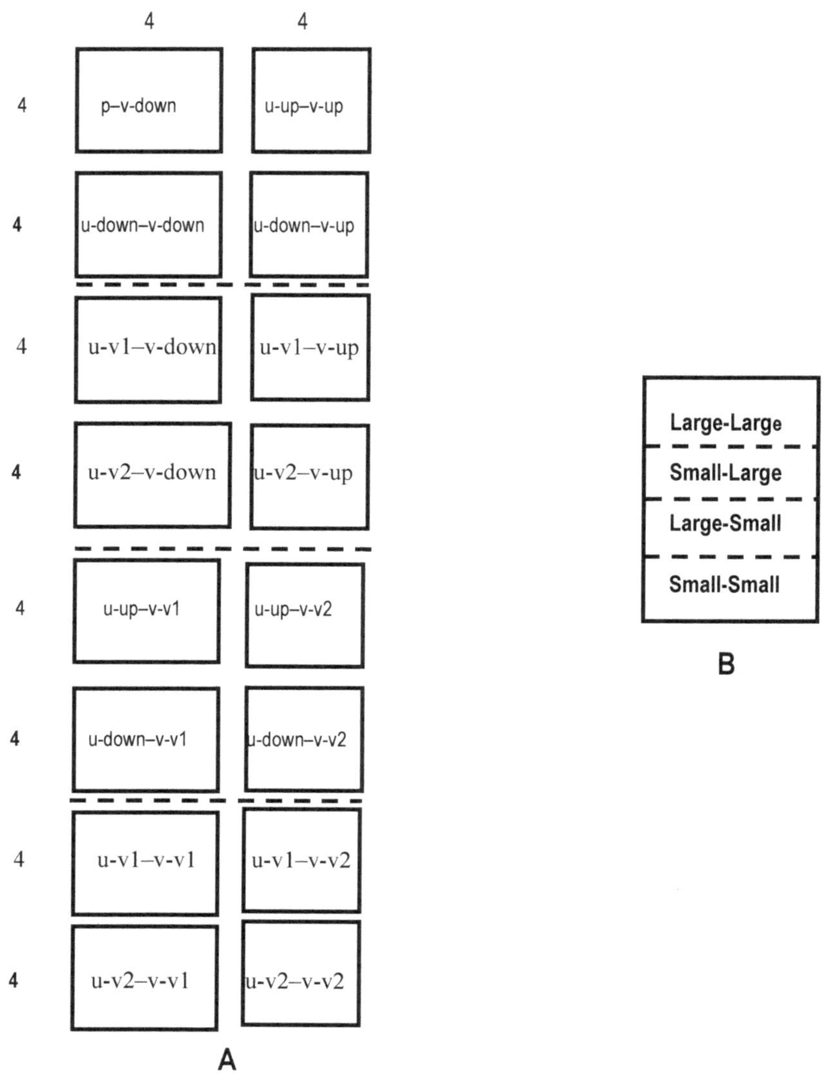

Figure 4.5. Fig. A: Block form of the 16 × 16 [U_{ba}] array arranged to a 32 × 8 form to display a layer structure similar to QUeST. Fig. B: The layers were grouped by sector type and ordered from "large" spinor components to "small" spinor components.

4.4 QUeST Dimension Array Inplied by 32 × 8 Spinor Array Form

The form of the 32 × 8 spinor array of Fig. 4.5 maps directly to the form of the QUeST dimension array. Fig. 4.6 shows the four layer dimension array's form. Each • represents a dimension in this Mathematical Picture Language depiction.

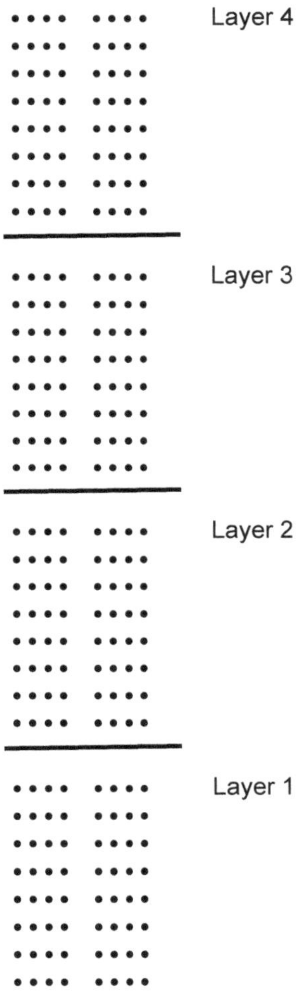

Figure 4.6. The 32 octonion dimensions QUeST array subdivided into 4 layers of 8 rows..Each layer will be seen to map to a block of fundamental group representations as shown in Figs. 4.7 and 4.8.

The layers of Fig. 4.6 have their dimensions allocated to various QUeST (UST) groups in Fig. 4.7. The breakdown of layers into blocks of 16 dimensions is shown in Fig. 4.8 for two out of the four layers.

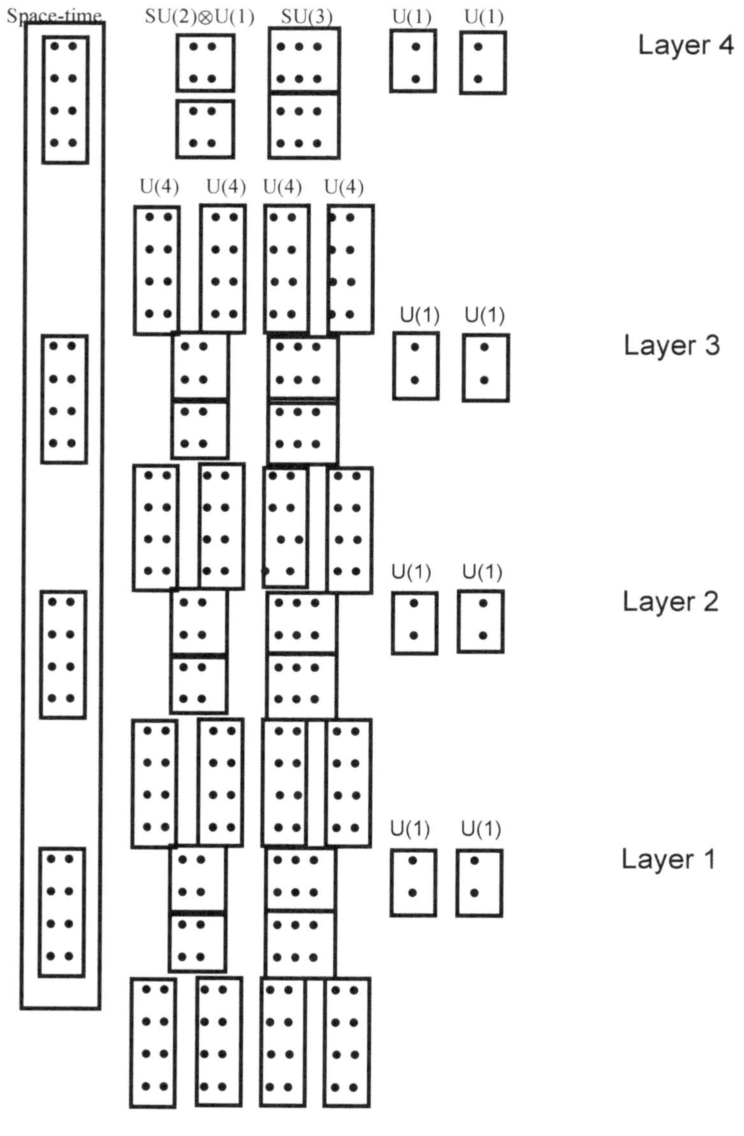

Figure 4.7. The four layers of QUeST internal symmetry groups (and space-time) for 32 dimension complex quaternion space. Note: each row has an 8 pebble • octonion. Note the left column of blocks combine to specify a 4 dimension octonion space-time. Note each layer has 64 dimensions.

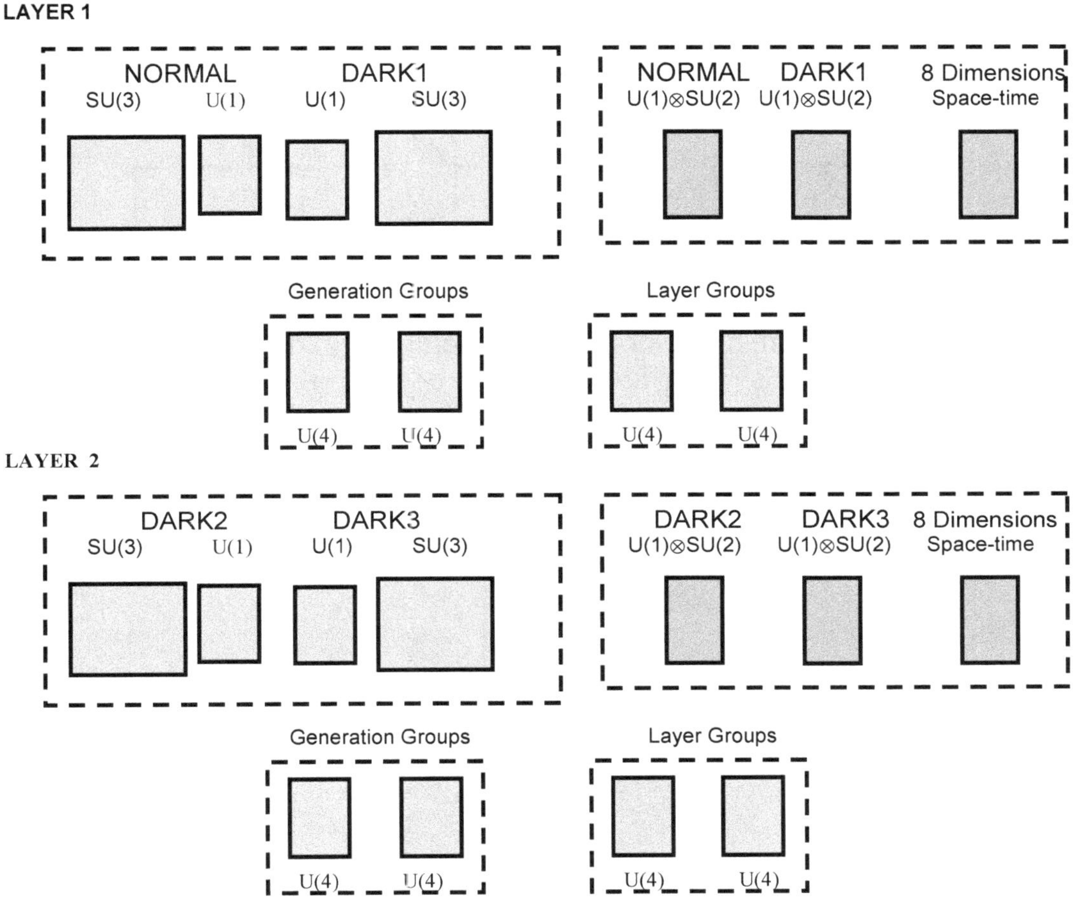

Figure 4.8. Sets of 16 dimension blocks for the first two layers (of four) of the 32 $\times$ 8 array. Each "dashed" block is a $4 \times 4 = 16$ array of dimensions. This set of 8 blocks contains the $8\times16 = 128$ dimensions of layer 1 and Layer 2.

4.5 Progression from BQUeST to UST

Fig. 4.9 shows the progression from BQUeST to UST. The structure of the Unified SuperStandard Theory (UST) reflects the structure of the spinor array constructed from the seed fermions. *Thus we have support at the level of Elementary Particles for the possibility of a fundamental BQUeST level residing in the eight dimension Magaverse space-time.*

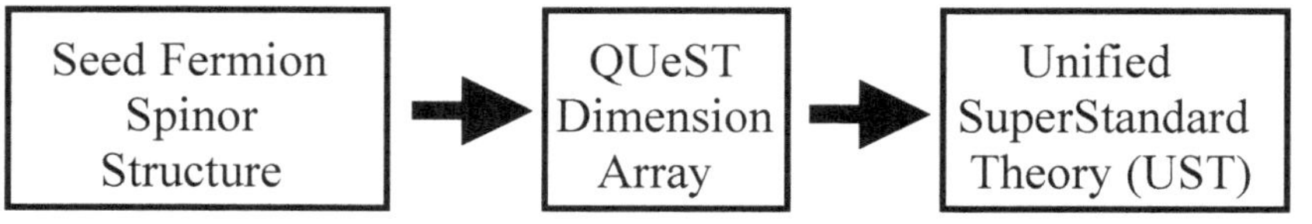

Figure 4.9. The seed fermion spinor array structure is mirrored by the QUeST dimension structure. This is evidence for the seed origin of QUeST via BQUeST.

5. QUeST Universe Structure & its Relation to UST

BQUeST, QUeST, and UST form a complete theory of elementary particles and Gravitation. In chapter 4 we derived QUeST from BQUeST. In this chapter we show that QUeST implies UST: the same internal symmetries, same fermions, vector bosons, and Higgs particles. *This chapter shows that starting from BQUeST with one fermion-antifermion annihilation in an 8 dimension Megaverse one can generate QUeST and show it is consistent with UST.*

The probability of a consistent match of octonion-based QUeST and logic-based UST would seem to be very low. The consistency suggests that we have the right complete theory of elementary particles and gravitation.

5.1 Internal Symmetries

QUeST has a 256 dimension space. This space furnishes a fundamental representation for U(128). This symmetry must be broken to obtain the internal symmetries and space-time of UST (and of the Standard Model within it). A pattern of symmetry breaking,[45,46] which leads to UST and the Standard Model is:

1. U(128) broken to U(32)⊗U(32)⊗U(32)⊗U(32) giving four layers
2. Each U(32) broken to a 32 dimension U(16) giving "Normal" and "Dark" sectors
3. Each U(16) broken to a 16 dimension U(8) giving subdivisions of "Normal" and "Dark"
4. Each U(8) broken to SU(2)⊗U(1)⊗SU(3)⊗U(1)⊗(4 dimension space-time part) or U(4)⊗U(4)

Figure 5.1. Pattern of symmetry breaking of QUeST U(128).

This pattern results in a "tiling" of the 256 dimension QUeST array with sixteen 16 dimension tiles composed of the groups:

$$\underline{\text{8 Tiles}} \qquad\qquad \underline{\text{8 Tiles}}$$
$$\text{SU(2)}\otimes\text{U(1)}\otimes\text{SU(3)}\otimes\text{U(1)}\otimes\text{(4 dimension space-time part)} \quad \text{or} \quad \text{U(4)}\otimes\text{U(4)} \qquad (5.1)$$

The space-time parts combine to give a four octonion space-time from which real four dimension space-time are extracted.

---

[45] The pattern presented here is an alternative view of the QUeST dimension array. The BQUeST derivation of Chapter 4 is fully consistent with it.

[46] Suggested by the derivation of QUeST in a two-step process: first using the one dimension fermion's four spinors (the fermion is in a 4-dimension external space) to boost 1-dimension BQUeST to a 4 × 4 array; then assuming each of the array components is a fermion using each's 4-spinors boost to obtain a 16 × 16 dimension QUeST array. See Blaha (2020i) for details.

The symmetry breaking of Fig. 5.1 lead to the internal symmetries of QUeST:[47]

$$[SU(2)\otimes U(1)\otimes SU(3)\otimes U(1)]^8\otimes U(4)^{16} \tag{5.2}$$

The corresponding layout of the QUeST dimensions appears in Fig. 5.2. There are four layers of internal symmetries in QUeST. Each layer has two SU(2)⊗U(1)⊗SU(3) groups,[48] two U(4) Generation groups, two U(4) Layer groups, and two U(1) Fermion groups (described later).[49]

[47] QUeST/UST also contains the U(4) Species group, which follows from the General Relativity of the space-time.

[48] SU(2)⊗U(1)⊗SU(3)⊗U(1) does not have an SU(5) covering group. The groups in QUeST do not support SU(5) characterizations.

[49] We have replaced the Dark U(2) group seen earlier in previous books with a pair of U(1) groups because we were better able to achieve the $4 \times 4 = 16$ dimensions pattern that seems required.

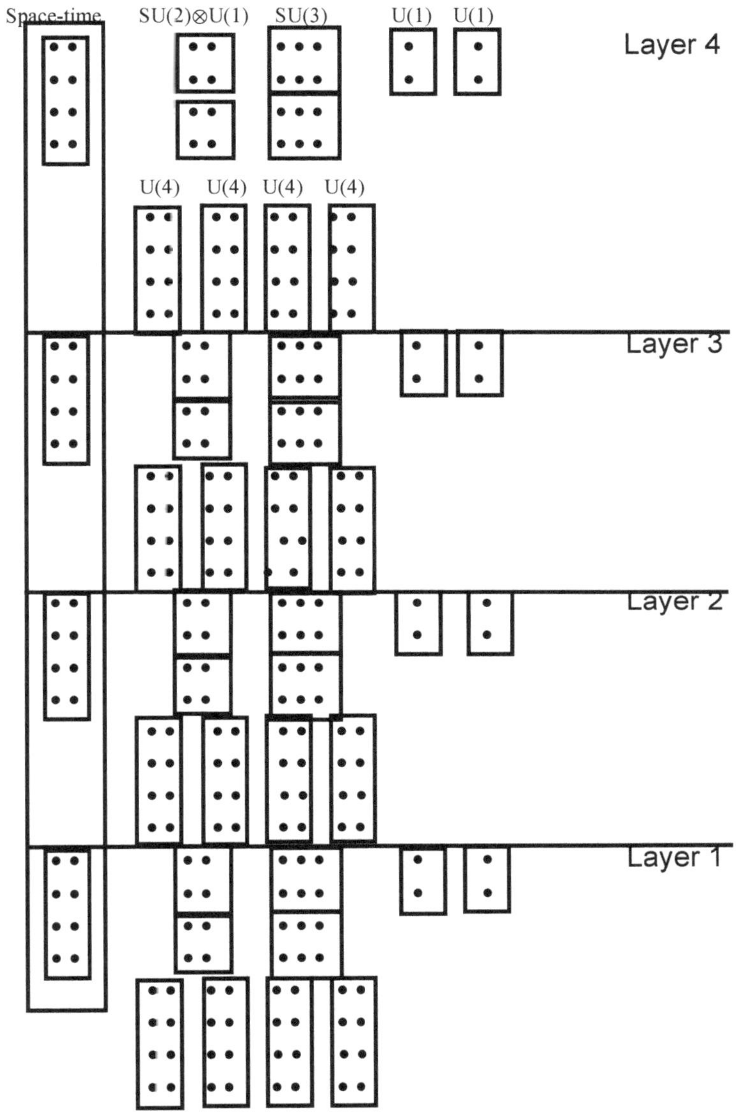

Figure 5.2. The four layers of QUeST internal symmetry groups (and space-time) for its 32 dimension octonion space. Note: each row has an 8 • octonion. Note the left column of blocks combine to specify a 4 dimension octonion space-time. Note each layer requires 64 dimensions.

Layers	NORMAL 4	NORMAL 4	DARK 4	DARK 4
4	SU(2)⊗U(1)⊗SU(3)⊗U(1) 4 Space-time Dimensions	Generation + Layer Groups	SU(2)⊗U(1)⊗SU(3)⊗U(1) 4 Space-time Dimensions	Generation + Layer Groups
4	SU(2)⊗U(1)⊗SU(3)⊗U(1) 4 Space-time Dimensions	Generation + Layer Groups	SU(2)⊗U(1)⊗SU(3)⊗U(1) 4 Space-time Dimensions	Generation + Layer Groups
4	SU(2)⊗U(1)⊗SU(3)⊗U(1) 4 Space-time Dimensions	Generation + Layer Groups	SU(2)⊗U(1)⊗SU(3)⊗U(1) 4 Space-time Dimensions	Generation + Layer Groups
4	SU(2)⊗U(1)⊗SU(3)⊗U(1) 4 Space-time Dimensions	Generation + Layer Groups	SU(2)⊗U(1)⊗SU(3)⊗U(1) 4 Space-time Dimensions	Generation + Layer Groups

Figure 5.3. Four layers of Internal Symmetry groups in QUeST. The groups in each layer are independent of those in other layers. The groups in each block of each layer are independent of those in the other blocks. Each block contains 16 dimensions. The block dimensions furnish fundamental representations for the groups listed. The entire set of blocks contains 256 dimensions. Each layer contains 64 dimensions. The first two columns are for the "Normal" sector. The last two columns are for the "Dark" sector (although most of the Normal sector is Dark observationally at present.) This Figure also holds for UST with the addition of Fermion groups.

Note: The columns labeled NORMAL in Figs. 5.3, 5.4 and later include three layers that are Dark in the sense that their interactions and matter are not as yet found experimentally. The columns labeled DARK above are Dark, now and in the future. DARK sectors are distinct from NORMAL sectors.

5.2 QUeST → UST Symmetries

The 16 tiles of Fig. 5.3 contain various symmetry groups and space-time. They are multiple copies of the same internal symmetries. However their coupling constants, particle masses, and symmetry breaking parameters will differ in general.[50]

---

[50] If the Blaha calculation of the ElectroWeak and Strong coupling constants based on vacuum polarization holds, then all copies of SU(2)⊗U(1)⊗SU(3) have the same coupling constant values. See Blaha (2020c) and earlier books. IF the Blaha vacuum polarization calculation of coupling constants *in all QUeST layers* does *not* hold, then Grand Unified Theory (GUT) unification is doubtful. IF the Blaha vacuum polarization calculation of coupling constants *in all QUeST layers holds*, then Grand Unified Theory (GUT) unification becomes possible.

We are familiar with part of one tile: three known generations of Normal fermions, the SU(2)⊗U(1)⊗SU(3) ElectroWeak and Strong interactions, and some masses, couplings and symmetry breaking parameters. The space-time is also known. The other tiles are completely unknown although our Quaternion-based theory suggests their likely form.

We now turn to mapping the above features of QUeST to UST. The QUeST symmetries listed in columns 1 and 3 of Fig. 5.3 do not appear in UST as originally known. We augmented that UST, which is derived from axioms, by including these U(1) symmetries. The space-time coordinates are separate in UST. UST has 3+1 dimensions separate from internal symmetry; QUeST unites all dimensions for internal symmetry and space-time—one of its motivating goals.

Otherwise the pattern of QUeST symmetries matches the pattern of UST. We used this *correspondence* to determine the QUeST symmetries of columns 1 and 3 in Fig. 5.3 with a U(1) Fermion group (discussed below) added to each of their tiles. Further we use the UST symmetries of columns 2 and 4 of Fig. 5.4 to determine the corresponding columns of QUeST. These choices are appropriate since the QUeST dimension array does not inherently determine the physical nature of QUeST groups. Instead we see a *Correspondence Principle* where UST determines the physics of the QUeST array. Consistency of the QUeST array groups with the UST groups is required. This section demonstrates consistency if Fermion U(1) groups are added to UST. Satisfyingly, the U(1) groups have a natural physical interpretation in UST.

We now consider the Generation, Layer and Fermion groups. These groups occur in UST. We map them back to QUeST using the previously mentioned Correspondence Principle..

5.3 U(4) Generation, U(4) Layer and U(1) Fermion Groups

The Generation, and Layer groups were introduced in UST. (See Blaha (2018e), 2020c) and earlier books.) Their basis is in number conservation laws such as Baryon conservation. The Fermion group, now introduced, is also based on conservation laws:

We now describe the Generation and Layer groups briefly. Blaha (2020c) provides a more detailed description as do earlier books. The Fermion group is described later.

5.4 The Generation Group

We define two particle number operators for normal up-quark particles and down-quark particles, B_{uq} and B_{dq}. Similarly we define two particle number operators for normal species "e" (electron) particles and species "ν" particles, B_e and B_ν. Similarly we define Dark matter equivalents:[51] B_{De}, $B_{D\nu}$, B_{Duq}, and B_{Ddq}.

[51] By analogy, we assume that there are four species of Dark matter: charged Dark leptons, neutral Dark leptons, Dark up-type quarks, and Dark down-type quarks. Thus we are led to the Dark particle numbers: Dark Baryon Numbers, and Dark Lepton Numbers shown above.

In the absence of interactions these fermion particle number operators are conserved. Each set are "diagonal" operators within a U(4) group. Thus we have a normal U(4) Generation Group and a Dark U(4) Generation group *for each layer.*.

On this basis we find there are four generations of each species in the normal and in the Dark matter sectors since U(4) groups have 4-dimension fundamental representations. One generation of normal fermions with large masses has not as yet been found in the lowest layer; three generations of normal fermions have been found. .

The broken symmetry gauge vector bosons of the Generation Group also have large masses. If the conservation of the fermion particle numbers is broken then we view it as a consequence of Generation Group symmetry breaking.

The Generation group of each layer generates interactions between the fermions of that layer. See Fig. C.5 of Appendix C for the interaction between fermions generated by the Generation group of layer 4.

5.5 The Layer Group

The set of particle number operators can be extended if we take account of the fourfold fermion generations. To further refine the set of particle number operators we temporarily neglect all interactions that would violate conservation laws for the set.

We therefore subdivide the above particle number set into four particle numbers per generation. For the ith generation we define

L_{ie} – The "e" species particle number for the ith generation
$L_{i\nu}$ – The ν species particle number for the ith generation
L_{iuq} – The up-quark species particle number for the ith generation
L_{idq} – The down-quark species particle number for the ith generation
L_{iDe} – The Dark "e" species particle number for the ith generation
$L_{iD\nu}$ – The Dark ν species particle number for the ith generation
L_{iDuq} – The Dark up-quark species particle number for the ith generation
L_{iDdq} – Dark down-quark species particle number for the ith generation

for each generation i = 1, 2, 3, 4. Individual fermions have positive $L_{ia} = +1$ values and anti-fermions have negative $L_{ia} = -1$ values for species a = 1, 2, 3, 4 (with the three color subspecies of quarks treated as part of one species.)

At this point we have four particle number operators for each generation. We define a group framework for each set of particle numbers. The simplest way is to assume that each generation consists of the four layers with the particles in each generation in a U(4) fundamental representation. Then each generation has a U(4) Layer group with the generation's four number operators (above) as its diagonal operators. We call this group the Layer Group of the ith generation L_{ia}. With four generations we obtain four U(4) Layer groups for normal matter. In addition there are four U(4) Dark Layer groups. See Fig. 5.3.

The consequence of this expansion of particle numbers and groups is that the set of fermions increases fourfold. We now have four layers, with each having four generations. Experimentally, we know of three generations of fermions—the lowest generations of the lowest level. The remaining generation and the three additional levels of fermions are of much higher mass and yet to be found.

See Blaha (2020c) and (2018e) for a detailed discussion of the Layer Groups. We note in passing that the symmetries of these number operators are badly broken. Yet the underlying group structure remains.

Note each Layer group provides an interaction among layers: the ith Layer group provides an interaction between the four i^{th} generations. See Fig. C.5 for the four Layer groups' interactions between fermions for the four generations. Layer groups provide the only interactions between layers.

5.6 The Fermion Groups

For the NORMAL sector of each layer there is a U(1) Fermion group.[52] There is a similar U(1) Fermion group for the DARK sector of each layer. See Fig. 5.3. These groups were introduced in QUeST. We have added them to UST since they have natural roles.

We will attribute them to fermion numbers conservation[53] for each layer and for NORMAL and DARK separately—one conservation law for total fermion number in each tile of columns 1 and 3 of Fig. 5.3. *In each of the columns 1 and 3 tiles fermion number equals the sum of Baryon number and Lepton number. It appears to be conserved in particle interactions. It can be conserved even if Baryon number conservation and Lepton number conservation were violated.*

The sum of all NORMAL Fermion numbers F_{tot} is conserved. The sum of all DARK Fermion numbers F_{Dtot} is conserved as well.

For the NORMAL and DARK sectors of each layer of fermions we define an additive Fermion number N_f that equals the number of fermions of that layer minus the number of antifermions of that layer. In an interaction where fermion number is conserved, the input N_f equals the output N_f. This conservation law, which resembles the baryon number conservation law, is broken by Layer group interactions. Thus the broken Fermion groups, like the Generation and Layer groups, are based on a conserved number. The gauge fields of the Fermion groups will be denoted Y^{μ} and $Y_D{}^{\mu}$ for the NORMAL and DARK sectors of each layer respectively.

UST with Fermion groups added follows from QUeST.

5.7 Particle Spectrums of QUeST: Fermions, and Vector Bosons

In this section we show that QUeST and UST have the same vector boson and fermion particle spectrums.

---

52 Earlier versions of QUeST in this year had a broken U(2) Dark symmetry group that gave interactions between Normal and Dark particles. Here we replace this group with U(1)⊗U(1) Fermion groups due to a tile by tile conserved fermion number. Appendix C uses the Dark U(2) group. The conent of that appendix can be modified to Fermion groups by replacing each Dark U(2) with U(1)⊗U(1).

53 Fermion number conservation differs from charge conservation since it applies to both charged and uncharged fermions.

5.7.1 Vector Bosons

An examination of Figs. 5.2 and 5.3, which now apply to both QUeST and UST, directly yields the spectrum of gauge vector bosons. This spectrum includes the known vector bosons of the Standard Model. The list of vector boson groups and interactions *for each of the four layers* is:

<u>NORMAL Gauge Groups</u>
SU(2)⊗U(1)⊗SU(3)⊗U(1)
Generation Group U(4)
Layer Group U(4)
<u>DARK Gauge Groups</u>
SU(2)⊗U(1)⊗SU(3)⊗U(1)
Generation Group U(4)
Layer Group U(4)

5.7.2 Fermions

Having the spectrum of gauge vector bosons the fundamental fermion spectrum also follows directly from the dimensions of the fundamental group representations. See Fig. C.5 for the QUeST and UST fermion spectrums.

The basis for the fermion spectrum is twofold:

1. There are four species of fermions due to the boosts of the Complex Lorentz group as shown in Blaha (2020c) and earlier books. These species are charged lepton, neutral lepton, up-type quark, and down-type quark.[54]
2. The leptons were seen to be color singlets. The quarks were seen to be Color triplets.
3. With these points in mind there are 8 fermions for each of the eight appearances of SU(2)⊗U(1)⊗SU(3)⊗U(1) in Fig. 5.3. See Fig. C.5.

5.8 Other QUeST-UST Topics

There are a number of important UST topics within QUeST that are presented in Blaha (2020c), (2018e) and other books by the author. They include:

1. Color Confinement
2. Gravitation Potential at Earthly, Galactic, and intergalactic Distances
3. Higgs Mechanism, Symmetry Breaking. Fermion and Gauge Boson Masses
4. Monads From factorization of Wave Functions to Eliminate Quantum Entanglement
5. Determination of α and Other Coupling Constants
6. Evidence for Faster-than-Light Particles and their Implications
7. Complex Gravitation and the Species group
8. An Equipartition Principle for Universe Matter and Energy Abundance

---

[54] While this was shown as early as 2007 by the author, it became very evident as a general form when it was shown to hold in the UTMOST Megaverse earlier in this year.

9. Θ rotations of dimensions and vector fields in UST analogous to that of Blaha (2018e)

6. UTMOST for the Megaverse

This chapter describes the quaternion octonion space[55] of UTMOST with the Fig. 1.4 table entry:

5 Quaternion Octonion (32)	Quaternion Octonion	32×32	8 complex octonion

The dimension array of the quaternion octonion space with the quaternion octonion coordinates (rows) is 32×32. It has 1024 dimensions.

6.1 UTMOST Space and Internal Symmetries

An octonion contains eight dimensions. A complex octonion contains sixteen dimensions. A *quaternion octonion* contains 32 dimensions. Fig. 6.1 depicts the 32 dimension complex octonion space as a 32×32 array of dimensions. It uses a "dot" or pebble • to represent a dimension. The dimensions of the space are not assigned physically until they are mapped to internal symmetry group fundamental representation dimensions and space-time dimensions. Rather than create a cumbersome coordinate-based notation we choose to use •'s.

• • • • • • • •   • • • • • • • •
• • • • • • • •   • • • • • • • •
• • • • • • • •   • • • • • • • •
• • • • • • • •   • • • • • • • •
• • • • • • • •   • • • • • • • •
...
• • • • • • • •   • • • • • • • •

Figure 6.1. The 32 quaternion octonion dimension UTMOST array. It is a 32×32 dimension array of •'s. It has 1024 dimensions. It is grouped into four layers in rows of eight.

The repetitive pattern of groups seen in QUeST leads us to assume that UTMOST has a similar repetitive pattern. We will use a four layer format for the 32×32 array of dimensions. Each layer consists of 8 rows of Fig. 6.1. Each layer can be put in a form analogous to Fig. C.4 (and to Fig. C.9). See Fig. 6.3.

[55] In our earlier books this year we used an equivalent 64 complex octonion dimension space for UTMOST. The difference between this form of space and the 64 complex octonion dimension space above is not physically meaningful at present. The difference will be physically meaningful if the masses of the fermion spectrum and the full pattern of symmetry breaking are determined. Then one can differentiate between the symmetry group spectrum and mass spectrums of the respective possibilities.

We map between dimensions and fundamental group representations. We use the maps in Table 6.2 to set up the group ↔ dimension map, bearing in mind the group representations of the Standard Model:

U(4)	↔	8 real dimensions
U(2)	↔	4 real dimensions
SU(3)	↔	6 real dimensions
U(1)⊗SU(2)	↔	4 real dimensions
U(1)	↔	2 real dimensions

Table 6.2. Map between fundamental representations and their dimensions.

Figs. 6.3 and 6.4 show the content of one UTMOST layer. The four layers of UTMOST are four copies of Fig. 6.3.[56] The separation of the set of dimensions is accomplished by following the procedure given earlier.[57]

Fig. 6.5 shows the 4 × 4 blocks in the four layers (each in two rows) of the 32 × 32 dimension UTMOST array. The 4 × 4 blocks are within the four block 8 × 8 sections for each pair: Normal+Dark1, Dark2+Dark3, Dark4+Dark5 and Dark6+Dark7. In total they form the 32 × 32 = 1024 UTMOST dimension array.

[56] The Layer groups are U(4) groups. They mix the generations of each of the top four layers, generation by generation, separately from the Layer groups mixing the lower four layers. This feature enables QUeST universes to be generated from either the top four layers or the lower four layers.

[57] The separation of the dimensions into the subgroup factors' representation can be implemented as group transformations and definitions using standard group theoretic methods. A more formal method for extracting the subgroup content of representations uses a symmetric group analysis of U(n) representation characters. See S. Blaha, J. Math. Phys. **10**, 2156 (1969) for a detailed discussion of this approach.

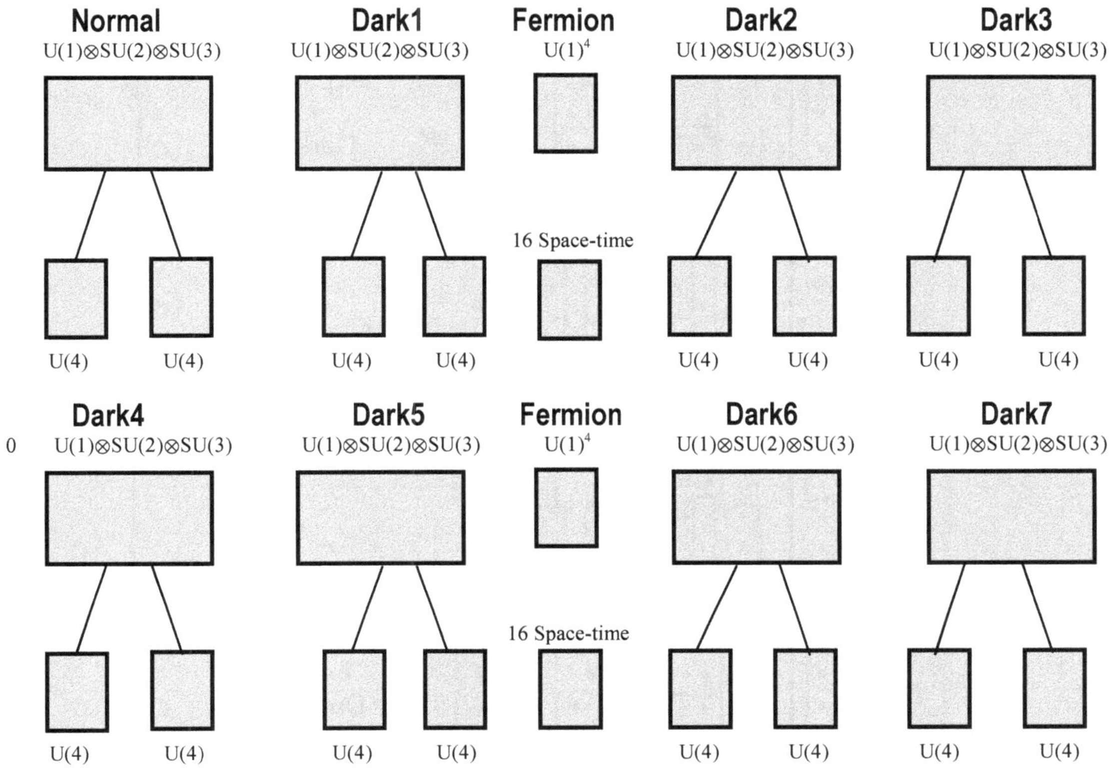

Figure 6.3. The internal symmetry groups of *one layer* (consisting of 8 rows in Fig. 6.1) of the four layers of 32×32 dimension UTMOST. The other three layers are copies of the this layer. Note the Fermion $U(1)^4$ groups.

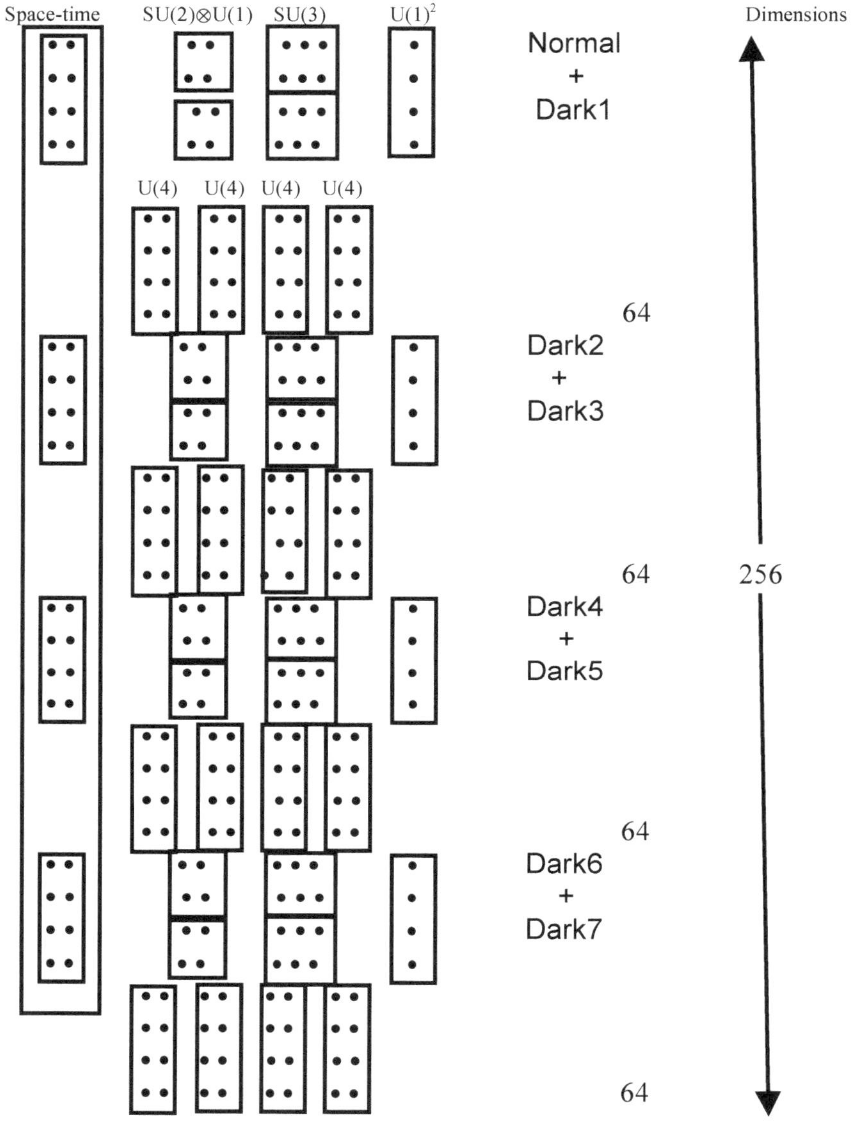

Figure 6.4 The *first* of the four layers of UTMOST dimensions with boxes around sets of dimensions for fundamental group representations.

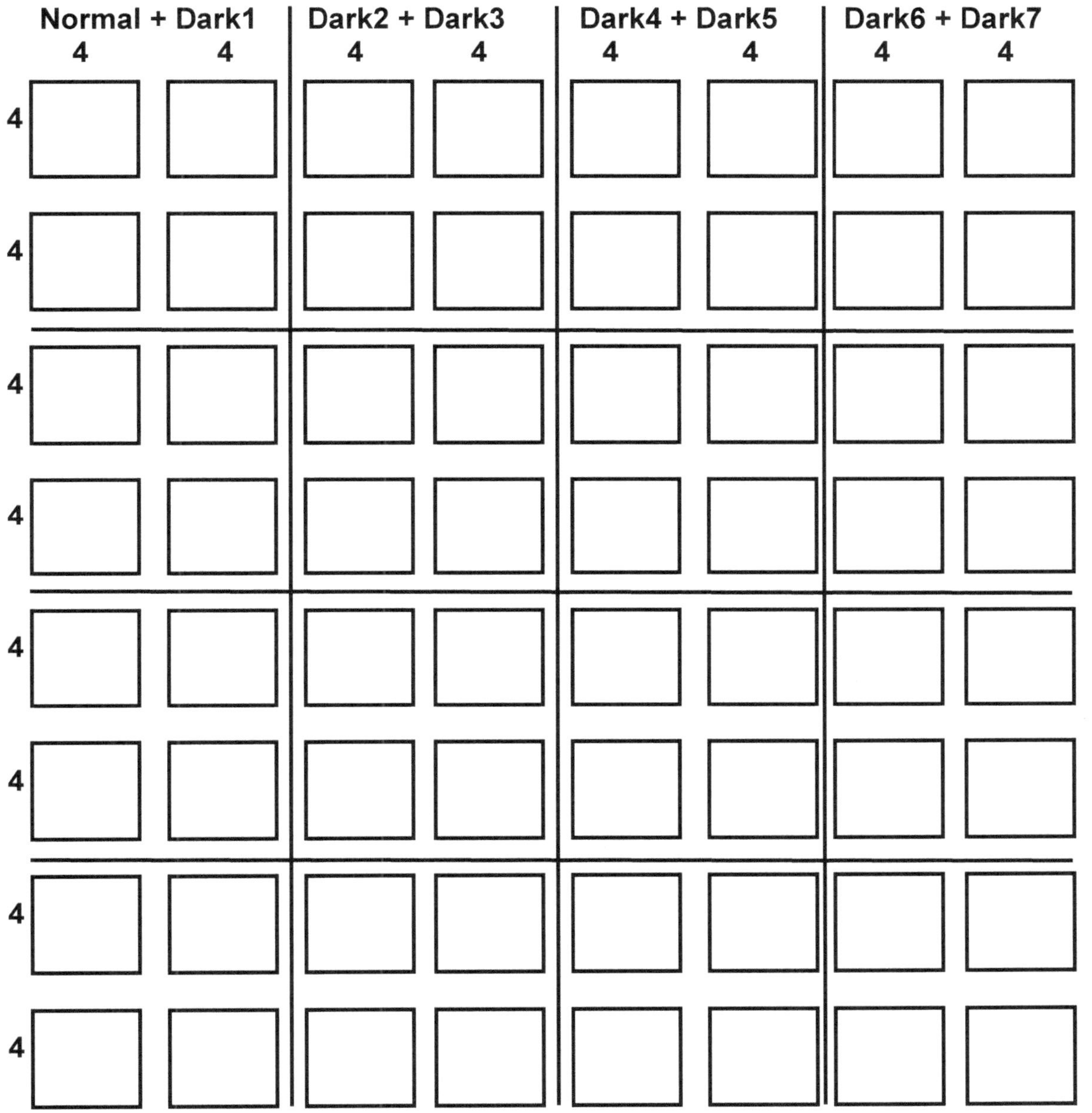

Figure 6.5. Four layers (each in two rows) in the 32×32 dimension UTMOST array composed of 4×4 blocks, which are within the four block 8×8 sections for each pair: Normal+Dark1, Dark2+Dark3, Dark4+Dark5 and Dark6+Dark7. In total they form the $32 \times 32 = 1024$ UTMOST dimension array.

The UTMOST dimension array is partitioned into $4 \times 4 = 16$ dimension blocks within $8 \times 8 = 64$ dimension blocks.[58] Fig. 6.5 displays the three types of 16 dimension blocks:

A. U(4)⊗U(4)
B. U(1)⊗SU(2)⊗U(1)⊗SU(2)
C. SU(3)⊗SU(3)⊗U(1)⊗U(1)⊗(Space-time).

These blocks are repeated throughout the four layers.

UTMOST has 32 blocks of type A, 16 blocks of type B, and 16 blocks of type C.

The QUeST dimension structure (Fig. 2.4) results.

6.2 UTMOST Symmetries

6.2.1 UTMOST Internal Symmetry Groups

The UTMOST internal symmetry group is

$$[SU(2)\otimes U(1)\otimes SU(3)]^{32}\otimes U(4)^{64}\otimes U(1)^{32} \tag{6.1}$$

6.2.2 UTMOST Space-Time

The space-time of UTMOST is 8 complex octonion dimensions.

6.2.3 UTMOST Fermion Spectrum

Given the form of the internal symmetries in UTMOST we can determine the fermions in the fundamental group representations as shown in Fig. 6.6. The set of Internal Symmetries determines the set of fermions in fundamental representations. The fermions are ordered by layer.

6.3 UTMOST Fermions

Given the form of the internal symmetries in UTMOST we can determine the fermions in the fundamental group representations as shown in Fig. 6.6.

[58] The $8 \times 8 = 64$ dimension blocks arise as the footprint of the urfermion-antiurfermion annihilation in the Maxiverse that generates an UTMOST Megaverse instance. See section 8.3.

UTMOST Fermion Array

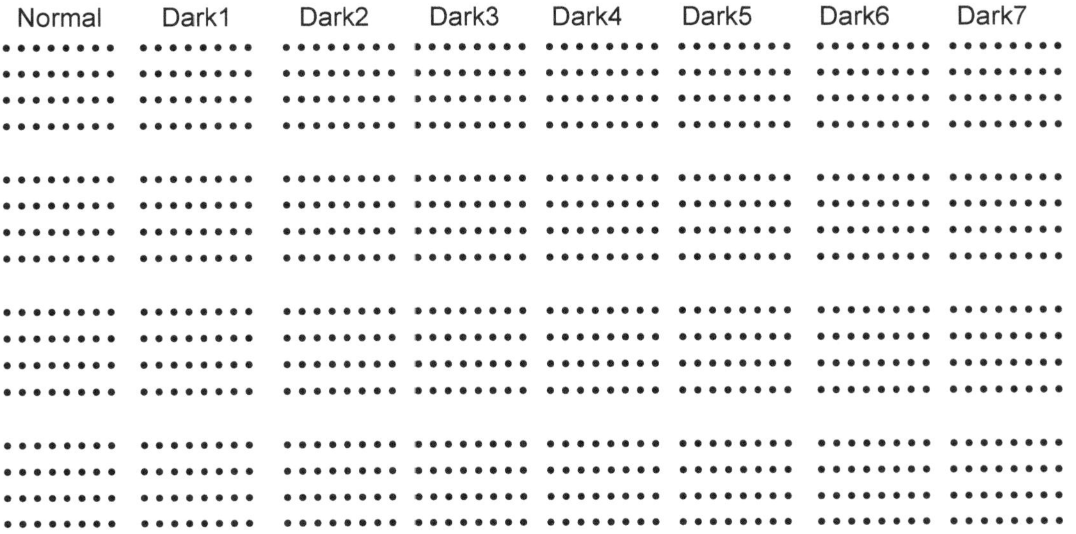

Figure 6.6. Spectrum of UTMOST fermions in a 16×64 format. Each fermion is represented by a •..Each set of eight •.'s represents a charged lepton, a neutral lepton, three up-type quarks, and three down-type quarks. There are eight sets of four species in four generations which are in turn in 4 layers. There are 1024 fundamental fermions taking account of quark triplets.

In chapter 2 we outlined possible patterns of subspaces of QUeST. One choice of pattern is based on 4 × 4 blocks of dimensions, assembled into 8 × 8 blocks of dimensions containing four 4 × 4 blocks, assembled in four layers.

Fig. 6.7 shows the possible implications of this arrangement for UTMOST fermions. The 4 × 4 fermion blocks contain either four generations of charged leptons and up-quarks, or four generations of neutral leptons and down-quarks.

The grouping of a lepton and three quarks in both cases creates a similarity to time and spatial coordinates respectively suggesting a broken Lorentz group-like structure or a possible SU(4) broken symmetry.

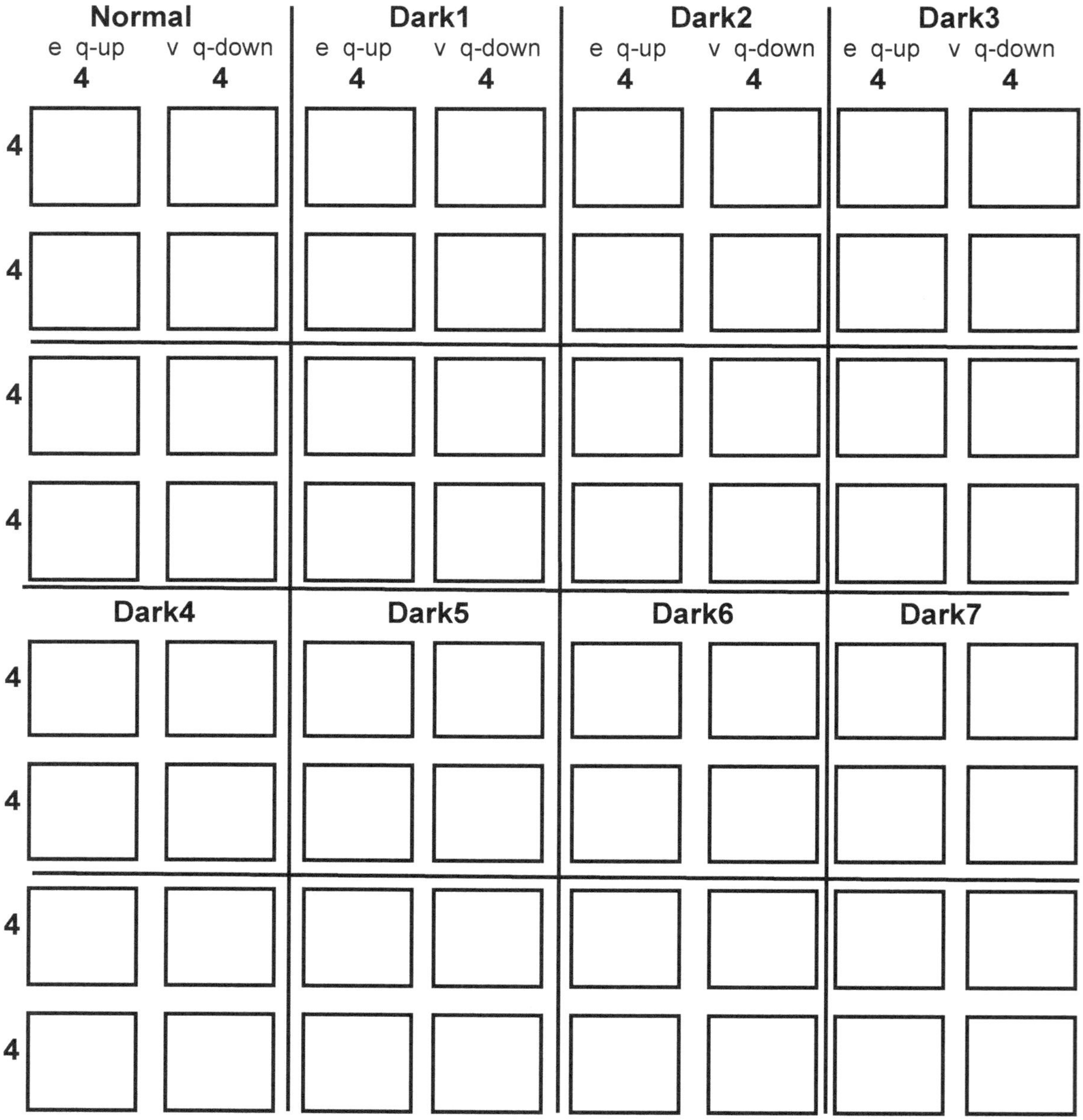

Figure 6.7. Block form of the 32 $\times$ 32 UTMOST fermion array with each row corresponding to *half of an UTMOST layer*. Thus 8 $\times$ ½ = 4 layers results. Each block contains four generations of fermions. The result is sixty-four 4 $\times$ 4 blocks. The label e q-up indicates a charged lepton – up-type quark pair, v q-down indicates a neutral lepton – down-type quark pair, and so on. *The form displayed here explains why generations come in fours.*

7. The BMOST Derivation of the UTMOST Megaverse

This chapter describes in detail the possible birth of the 1024 dimension UTMOST Megaverse instance from a fermion-antifermion annihilation in a separate 10 dimension space.[59] It describes the BMOST theory developed in 2020 in earlier books by the author.

BMOST[60] assumes a Megaverse begins from the annihilation of a fermion-antifermion pair in a 10 dimension space-time.

The derivation of the 32 × 32 = 1024 dimension UTMOST array requires a 10 dimension seed fermion, which we call the *urfermion.*[61] The 32 spinor components of a 10 dimension urfermion are used to generate the 32 × 32 dimension array of UTMOST.

We propose a picture of a 10 dimension fermion —the urfermion—that acts as the seed of the consequent Megaverse. The evolution (perhaps instantaneous) of the Megaverse begins with seed-antiseed annihilation at great energy to become a 1024 dimension UTMOST Megaverse. *An urfermion, which is off shell, has 32 independent spinor components that lead to the 1024 dimensions of the UTMOST Megaverse.* It generates internal symmetries and the space-time of UTMOST. Subsequently universes appear within the Megaverse as described in chapter 3.

The derivation parallels that of universe birth in the eight dimension Megaverse. Appendix E describes features of the UTMOST space.

7.1 BQUeST Origin

We begin with the assumption that an UTMOST Megaverse originates in a 10 dimension space..[62] The 10 dimension seed urfermion has 32 spinor components. (There are 1024 Dirac matrices, denoted G_{ab} for a, b = 1, …, 32, and ten 32×32 Dirac γ matrices.)

We picture an off shell seed urfermion and an off shell seed antiurfermion appearing as fluctuations in the 10 dimension space vacuum. They "annihilate" to create a Megaverse particle.[63] The Megaverse particle expands to become the Megaverse—the home of universes.

---

[59] The prevalence of 10 dimension spaces in SuperString theories raises the possibility of a relation of the Megaverse to a SuperString space.

[60] See Blaha (2020d) and later books.

[61] We use the Indoeuropean prefix ur- to signify original (or earliest). The author used this prefix in the 1970s in published Physics Letters B papers on quarks and leptons including discrete scaling fermion masses.

[62] The UTMOST Megaverse is described in Appendix E.

[63] Appendix D provides experimental and theoretical evidence for a particle view of universes. Comments by DeWitt and others on quantum universes support this view. A quantum Megaverse (required for consistency) must be aparticle too.

We will show the 32 independent spinor components of the 10 dimension off shell seed urfermion and of the off shell anti-seed urfermion generate the 32 × 32 dimension array of QUeST.[64]

7.2 A Model Dynamics of the Seed – Anti-Seed Generation of the Megaverse

The seed urfermion must have a dynamics that enables it to generate the 1024 dimension array. In this section we describe a model dynamics for the annihilation of a seed–antiseed state into a scalar Megaverse particle. The model must have certain features that accomplish the goal:

1. Since the target dimension array is not symmetric the PseudoQuantum formulation[65] of Quantum Field Theory will be seen to be required.

2. Since the quantum field theory is in 10 dimensions we must use Two-Tier coordinates,[66] which eliminate all divergences in perturbation theory using exponential damping of all integrations.

These extensions of quantum field theory were developed by the author in the 1970's and the early 2000's. They are both described in Appendix A together with references to earlier papers and work.

We define a model lagrangian for the seed urfermion sector. The seed urfermion is represented by *two* quantum fields[67] ψ_1 and ψ_2. The scalar universe particle has two quantum fields A_{1ab} and A_{2ab} with indices a and b.

The lagrangian terms are

$$\mathscr{L} = F^{1\mu\nu} F^2{}_{\mu\nu} + \bar{\psi}_2 (i\gamma\cdot\partial - m)\psi_1 + \bar{\psi}_1 (i\gamma\cdot\partial - m)\psi_2 - e_0\bar{\psi}_2 A_2\psi_1 - \bar{\psi}_1 A_1\psi_2 \qquad (7.1)$$

where m is the mass of the seed urfermion, and where $A_i = A_{iab}G_{ab}$ (summed over a and b) are the Megaverse fields.

To insure *finite* perturbation series results (in 10 dimensions) the fields are all functions of Two-Tier coordinates X having a vector quantum field $Y^\mu(y)$:

$$X^\mu(y) = y^\mu + i\, Y^\mu(y)/M_c^2 \qquad (7.2)$$

---

[64] The 10 dimension space may be a 10 complex octonion space-time. We choose to extract the real coordinate of each complex octonion 16-tuple to construct a real 10 dimension space-time within which seed urfermion dynamics occurs. This is analogous to the case of complex 4 dimension space-time being restricted to real complex space-time in conventional quantum field theory. Streater (2000) maintains complex space-time is needed for axiomatic quantum field theory.

[65] S. Blaha, Il Nuovo Cimento **49A**, 35 (1979) and Appendix A.

[66] See Appendix A.

[67] We use a PseudoQuantum Electromagnetic-like pair of fields also. The electromagnetic-like particle is a model of the universe. See eqs. 61 – 90 in Appendix 1-C of Blaha (2020j). The Two-Tier formulation eliminates all perturbation theory divergences in four dimensions including the notorious fermion triangle divergence. It also eliminates all perturbation theory divergences in eight dimensions.

with the Y^μ index $\mu = 0, 1, \ldots, 9$, and where M_c is an extremely large mass.

7.2.1 U(32) Species Group Symmetry for Dirac Matrix Indices

The lagrangian terms (plus additional universe terms) are invariant under a U(32) Species symmetry group. Appendix B describes the analogous U(4) Species[68] symmetry for four dimension fermions that we found for UST.

Appendix B shows the form of the (vector field) trial lagrangian terms for establishing the Species group of UST in conventional quantum field theory:

$$\bar{\psi}'(x)[i\gamma_\mu'(x)(\partial/\partial x_\mu - ig_8 A'_{Sk}{}^\mu(x)G_{Sk}) - m]\psi'(x) \tag{51.2}$$

where the matrices G_{Sk} are the 16 Dirac matrices of 4 dimension space, where k ranges from 1 through 16, and where the $A'_{Sk}{}^\mu$ fields.are 16 gauge fields for the Species group.

In developing the U(4) Species group symmetry for four dimension UST we related the indices of the Dirac matrices $[G_{Sk}]_{ab}$ where a, b = 1, 2, 3, 4 to the space-time coordinates using vierbein transformations. In this regard Appendix B states:

> A complex transformation of types II and III in Fig. 50.1 has the form:
>
> $$U(x'')^\mu{}_\nu = w^\mu{}_a(x'')[\exp(i\textstyle\sum_k \Phi_k(x'')\tau_k)]^a{}_b\, 1^b{}_\nu(x'')$$
> $$U^{-1}(x'')^\mu{}_\nu = w^\mu{}_a(x'')[\exp(-i\textstyle\sum_k \Phi_k(x'')\tau_k)]^a{}_b\, 1^b{}_\nu(x'')$$
>
> where τ_k is a U(4) generator matrix. Its infinitesimal transformation is approximately
>
> $$U(x'')^\nu{}_\beta \approx \delta^\nu{}_\beta + i\textstyle\sum_k \Phi_k(x'')[\tau_k]^\nu{}_\beta \tag{50.12}$$
> $$U^{-1}(x'')^\nu{}_\beta \approx \delta^\nu{}_\beta - i\textstyle\sum_k \Phi_k(x'')[\tau_k]^\nu{}_\beta$$
>
> using the *vierbein* flat space-time limits
>
> $$w^\mu{}_a(x'') \approx \delta^\mu{}_a \tag{50.12a}$$
> $$1^b{}_\nu(x'') \approx \delta^b{}_\nu$$
>
> where
>
> $$\Phi_k(x) = \int^x dy_\lambda\, A_{Rk}{}^\lambda(y) \tag{50.13}$$
>
> Then
>
> $$\Gamma_R{}^\sigma{}_{\lambda\mu} = -\tfrac{1}{2}i\{\textstyle\sum_k A_{Rk}(x'')_\mu[\tau_k]^\sigma{}_\lambda + \sum_k A_{Rk}(x'')_\lambda[\tau_k]^\sigma{}_\mu\} \tag{50.14}$$
> $$= A_R{}^\sigma{}_{\mu\lambda} + A_R{}^\sigma{}_{\lambda\mu}$$
>
> (summed over k) with the matrix $A_R{}^\sigma{}_{\mu\lambda}$ given by

68 The U(4) Species symmetry group is part of UST. It originates in the Complex General Reloativity of four dimension coordinates as shown in Appendix A and Blaha (2020c) as well as earlier books.

$$A_R{}^{\sigma}{}_{\mu\lambda} = -\tfrac{1}{2}i\sum_k A_{Rk\mu}[\tau_k]^{\sigma}{}_{\lambda} \qquad (50.15)$$

with $A_R{}^{\sigma}{}_{\mu\lambda}$ transformable to matrix row and column numbers

$$A_{R_{flat}}{}^{\mu a}{}_{b} = A_{R_{flatk}}{}^{\mu}[\tau_k]^{\sigma}{}_{\lambda}\delta_{\sigma}{}^{a}\delta^{\lambda}{}_{b} \qquad (7.3)$$

using the flat space-time vierbein values, and so $A_{R_{flat}}{}^{a}{}_{\mu b}$ may be written in matrix form as

$$A_{R_{flat}\mu} = -\tfrac{1}{2}i\sum_k A_{R_{flatk}\mu}\tau_k \qquad (50.16)$$

In the flat space-time limit the $A_{Rk}{}^{\lambda}(y)$ becomes the Coordinate Species group U(4) gauge fields $A_{R_{flatk}}{}^{\lambda}(y)$.

The above extract shows that space-time dimensions can be directly related to matrix indices. We will use this feature to map the indices of the Megaverse fields A_{iab} to internal dimensions of the Megaverse particle. The Megaverse dimensions, which number 1024 in the 10 dimension space-time, become the 1024 dimensions in UTMOST. (It should be remembered that the UTMOST Megaverse has 1024 dimensions —much more than the 256 dimensions of QUeST.)

7.3 The Interaction Generating a Megaverse Particle

The diagram for the creation process is in Fig. 7.1. A seed urfermion and an antiseed urfermion annihilate to create a Megaverse particle.

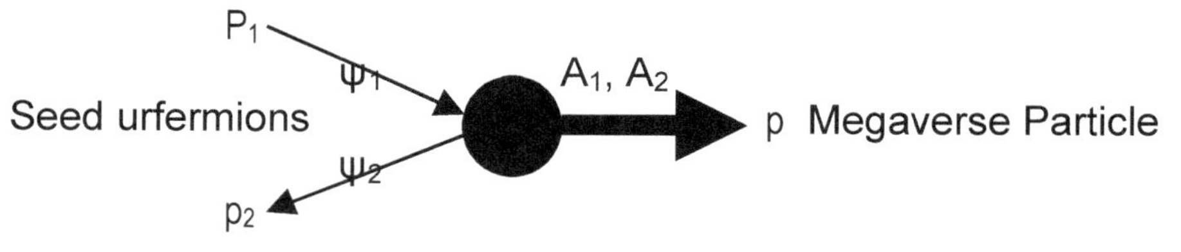

Figure 7.1. Diagram for the transition of seed - antiseed urfermions to a Megaverse particle.

The lowest order S-matrix element for this interaction has the form:[69]

$$S = \int d^{10}x\, [\overline{\psi}_2 A_2 \psi_1 + \overline{\psi}_1 A_1 \psi_2] \qquad (7.4)$$

After evaluating the integration we find S has term of the form

[69] Although the Maxiverse has 10 quaternion octonion space-time dimensions we can reduce it to 10 real-valued dimensions.

$$\bar{v}_{2a}(p_2)\ A_{1ab}(p)\ u_{1b}(p_1) \tag{7.5}$$

Rearranging the expression we obtain the trace expression

$$\text{Tr}\ \ u_{1b}(p_1)\bar{v}_{2a}\ (p_2)\ A_{1ab}(p) = U_{ba}A_{1ab}(p) \tag{7.6}$$

where U_{ba} is an array composed of u and $\bar{v}$.

Note the interaction can only take place for far off shell seed and antiseed fermions. Then the 32 spinor components of each fermion are independent and can be treated as independent.

Megaverse production (Fig. 7.1) conserves momentum. The diagram has two interesting cases:

1. If A_{1ab} and A_{2ab} are both proportional to a Kronecker δ_{ab} then the S matrix expression of eq. 7.4 simplifies to the scalar A_i case.

2. If A_{1ab} and A_{2ab} are not proportional to a Kronecker δ_{ab} then the S matrix expression can be expressed as a sum of terms for each value of a and b:

$$S = \Sigma\ S_{ab} \tag{7.7}$$

where

$$S_{ab} = \int d^8x\ [\bar{\psi}_{2a}A_{2ab}\psi_{1b} + \bar{\psi}_{1a}A_{1ab}\psi_{2b}] \tag{7.8}$$

The second expression eq. 7.8 suggests each "ab" pair of values marks a channel from the urfermions to the Megaverse particle. Thus we can view the Megaverse particle as receiving 1024 "streams" from the fermion-antifermion pair.

In section 7.4 below we suggest the streams "amalgamate" to form a 1024 dimension Megaverse.

7.4 U(16) Symmetry of A_i

The lagrangian in eq. 7.1 has a U(32) symmetry that is analogous to the 4 dimension U(4) Species group described in section 51.1 of Appendix B. This U(32) symmetry exists in spinor space. The matrix, $[U_{ba}]$ with its 32-spinor indices a and b, has a symmetry under the U(32) transformation:

$$U[U_{ba}]U^{-1} \tag{7.9}$$

U(32) symmetry requires the A_i factors in eqs. 7.5 and 7.6 to also have a U(32) transformation of the form:

$$A_i' = UA_iU^{-1} \tag{7.9a}$$

This symmetry will be seen to impact on the form of the UTMOST dimension array.in chapter 8.

7.5 Indices of $[A_{1ab}]$ Transformed to Internal Dimensions

Eq. 7.5 shows the indices of the array, $[A_{1ab}]$, have $32 \cdot 32 = 1024$ values due to the 32×32 independent values of the products: $u_{1b}(p_1)\overline{v}_{2a}(p_2)$. Eq. 7.3 above shows that indices can be transformed to coordinates.

Therefore we can take the array of indices [ab] and transform them to coordinates using a 10-space vierbein.[70] We can take the coordinates that emerge as each corresponding to a one dimension line (a stream?) within the Megaverse particle. Subsequently, or simultaneously, after creation the energy-momentum within the streams it can be distributed/flow into the 8 dimension space-time of the UTMOST Megaverse. The internal energy-momentum must sum to zero for momentum conservation. So there will be patches of positive and negative energies. (Negative energies may be absorbed by a vacuum rearrangement.)

The independence of the spinor components of each of the two seed urfermion fields guarantees the elements of the array of indices form an array of independent dimensions.

7.6 The Difference between Indices and Internal Dimensions

The array $[A_{1ab}]$ is an array with indices. We now consider the relation of indices and internal dimensions of an entity. First we consider the case of a cube, which can be viewed as an entity with three indices. Its external indices can be taken to label the length, width and height of the cube. Within the cube we can establish a coordinate system with three dimensions labeling points within it. Thus external indices can be viewed as mapping to internal dimensions within.

Perhaps a better example is a Black Hole. It exists in 4-dimension space. Within its event horizon the coordinates of the external space can be continued. It is also possible to replace them with an internal coordinate system of four dimensions. This possibility could be supported by the role of radial dimension externally becoming effectively a time coordinate internally.

We thus conclude the indices of A_{iab} can be viewed as specifying coordinates (*dimensions)* within A_i. We have mapped the seed urfermion to a $32 \times 32 = 1024$ dimension array. This array serves as the dimensions of the UTMOST Megaverse space.

7.7 Origin of the Megaverse

We can then envision the possibility that the Megaverse (multiverse) started by the annihilation of a seed urfermion and antiurfermion in a 10-dimension space. It "acquired" the 1024 dimensions of the UTMOST Megaverse to form a space that includes the 7+1 dimension space-time at the Megaverse creation point as well as

[70] The Megaverse undoubtedly has an 8-dimension General Relativity. For small neighborhoods of a point the vierbein becomes the identity as eq. 50.12a above shows.

internal symmetries. This process could occur, as it likely does, instantaneously. The seed urfermion-antirfermion are necessarily in a 10-dimension space as we showed in previous books

7.8 Nature of the Seed Urfermion and Antiurfermion

The type of the seed urfermion and antiurfermion in 10-dimension space can be partly ascertained. In Blaha (2020e) we showed that QUeST and UTMOST fermions occur in four species: charged leptons, neutral leptons, up-type quarks, and down-type quarks. Examples are e, ν_e, u, and d.

Similarly, there are four layers of urfermions in 10-dimension space if it supports a 10-dimension Special Relativity (and General Relativity). Each layers has sets of four generations of four species. A reasonable possibility is the seed urfermion is an *electron-type fermion* with the lowest mass of all 10-dimension space urfermions.

Thus the nature of the fundamental urfermion may be partly known.

8. Connection of Seed UrFermion Structure to UTMOST Dimensions Structure

Decisive experimental evidence for the origin of the Megaverse in seed urfermion-antiurfermion annihilation may never be found. There is a possibility that theoretical support for our proposed origin of the Megaverse may arise. It is based on the possibility that the structure of the UTMOST dimension array may have the "footprint" of the form of the seed urfermion-antiurfermion prior to annihilation.

The only reasonable long-term hope is that the seed urfermion-antiurfermion annihilation that produces an UTMOST Megaverse has features that appear to impress themselves on Megaverse structure. This chapter explores that possibility. The procedure is analogous to the BQUeST footprint discussion of chapter 4.

8.1 Structure of 32-Spinor u and v

The thirty-two 32-spinors of a 10 dimension urfermion have the general form in Fig. 8.1.

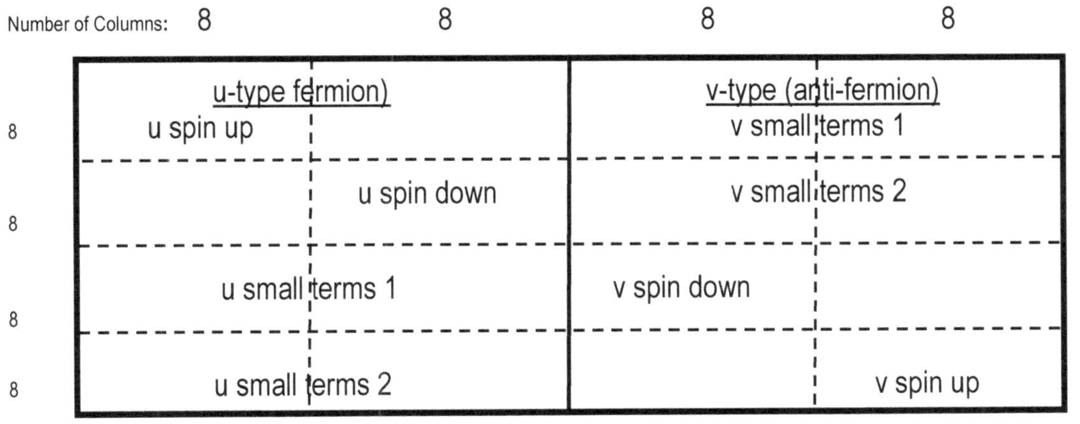

Figure 8.1. The 32 spinors of a 10 dimension spinor. Each spinor has 32 rows.

8.2 Compression of u and v into Composite Spinors

We can combine the sixteen u-type spinors into a composite spinor to exhibit their structure. Similarly, we can combine the sixteen v-type spinors into a composite spinor. Fig. 8.2 shows the forms of the *composite* u-type and v-type spinors.

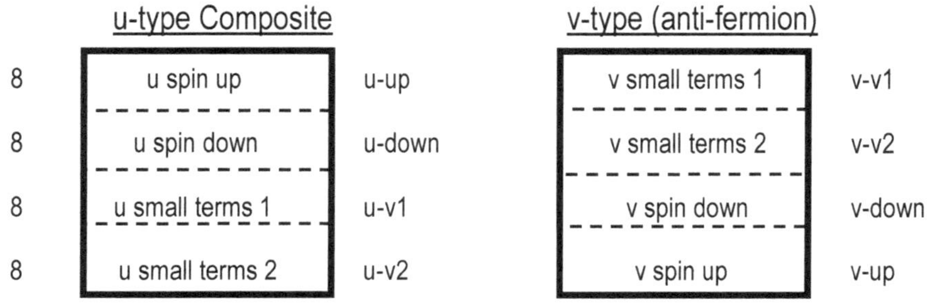

Figure 8.2. Composite u-type and v-type spinors illustrating the structure of each type of spinor. Each spinor is subdivided into 4 sections of 8 components. The third and fifth columns are symbols for each sector.

The outer product of spinor factors in eq. 7.6 is $u\bar{v}$. Fig. 8.3 contains the form implied by the composite spinors.

Since the annihilation process for the unpolarized seed urfermion-antiurfermion pair requires a sum over spin, *the four columns in Fig. 8.3 become 32 columns due to the spin summation* A 10 dimension fermion has sixteen up-spin values and sixteen down spin values.

u-up–v-v1	u-up–v-v2	u-up–v-down	u-up–v-up
u-up–v-v1	u-up–v-v2	u-up–v-down	u-up–v-up
u-up–v-v1	u-up–v-v2	u-up–v-down	u-up–v-up
u-up–v-v1	u-up–v-v2	u-up–v-down	u-up–v-up
u-up–v-v1	u-up–v-v2	u-up–v-down	u-up–v-up
u-up–v-v1	u-up–v-v2	u-up–v-down	u-up–v-up
u-up–v-v1	u-up–v-v2	u-up–v-down	u-up–v-up
u-up–v-v1	u-up–v-v2	u-up–v-down	u-up–v-up
u-down–v-v1	u-down–v-v2	u-down–v-down	u-down–v-up
u-down–v-v1	u-down–v-v2	u-down–v-down	u-down–v-up
u-down–v-v1	u-down–v-v2	u-down–v-down	u-down–v-up
u-down–v-v1	u-down–v-v2	u-down–v-down	u-down–v-up
u-down–v-v1	u-down–v-v2	u-down–v-down	u-down–v-up
u-down–v-v1	u-down–v-v2	u-down–v-down	u-down–v-up
u-down–v-v1	u-down–v-v2	u-down–v-down	u-down–v-up
u-down–v-v1	u-down–v-v2	u-down–v-down	u-down–v-up
u-v1–v-v1	u-v1–v-v2	u-v1–v-down	u-v1–v-up
u-v1–v-v1	u-v1–v-v2	u-v1–v-down	u-v1–v-up
u-v1–v-v1	u-v1–v-v2	u-v1–v-down	u-v1–v-up
u-v1–v-v1	u-v1–v-v2	u-v1–v-down	u-v1–v-up
u-v1–v-v1	u-v1–v-v2	u-v1–v-down	u-v1–v-up
u-v1–v-v1	u-v1–v-v2	u-v1–v-down	u-v1–v-up
u-v1–v-v1	u-v1–v-v2	u-v1–v-down	u-v1–v-up
u-v1–v-v1	u-v1–v-v2	u-v1–v-down	u-v1–v-up
u-v2–v-v1	u-v2–v-v2	u-v2–v-down	u-v2–v-up
u-v2–v-v1	u-v2–v-v2	u-v2–v-down	u-v2–v-up
u-v2–v-v1	u-v2–v-v2	u-v2–v-down	u-v2–v-up
u-v2–v-v1	u-v2–v-v2	u-v2–v-down	u-v2–v-up
u-v2–v-v1	u-v2–v-v2	u-v2–v-down	u-v2–v-up
u-v2–v-v1	u-v2–v-v2	u-v2–v-down	u-v2–v-up
u-v2–v-v1	u-v2–v-v2	u-v2–v-down	u-v2–v-up
u-v2–v-v1	u-v2–v-v2	u-v2–v-down	u-v2–v-up

Figure 8.3. Outer 32×32 product array $[U_{ba}]$ (eq. 7.9) of the composite u-type and v-type spinors illustrating the structure of the outer product array of uv's. Each column represents 8 spinor columns, making 32 columns in all.

8.3 Block Form of [U_{ba}] Array

We now turn to partitioning the [U_{ba}] and identifying the blocks of the UTMOST array implied by the partition. Each column in Fig. 8.3 becomes eight columns in [U_{ba}] extended by the summation over the spin of the unpolarized seed fermions. The pattern of array entries is displayed in Fig. 8.4.

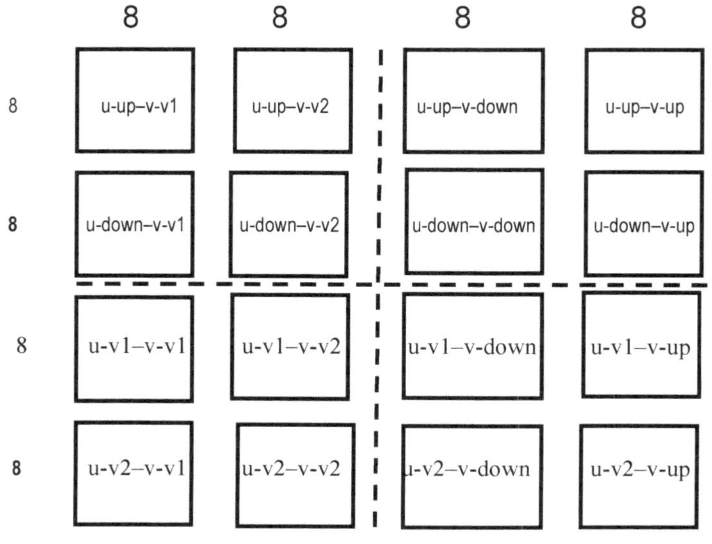

Figure 8.4. Block form of the 32 × 32 [U_{ba}] array. This is also the form of the UTMOST dimension array of 1024 dimensions. The sets of four blocks are divided by dashed lines that specify 256 dimension sections. These sections map to layers in UTMOST as shown in Figs. E.3, E.4 and E.5 for one of the four UTMOST layers. Each UTMOST layer is equivalent to a QUeST 256 dimension array.

The 8 × 8 blocks in the array match the 8 × 8 blocks (which further decompose into 4 × 4 blocks) that we found in earlier books such as Blaha (2020j). Figs. E.5 and E.7 show the 8 × 8 block structure of UTMOST dimensions *and* fermions. Each set of four blocks represents one of the four layers of UTMOST.

Fig. 8.5 shows the layout of groups in a 16 × 16 four 8 × 8 block section of Fig. 8.4.

The 8 × 8 = 64 dimension blocks show the correspondence of the seed urfermion-antiurfermion spinor structure to the UTMOST dimension structure. Given the structure of the Megaverse one could verify the possible origin of the Megaverse in BMOST urfermions in 10 space-time dimensions.

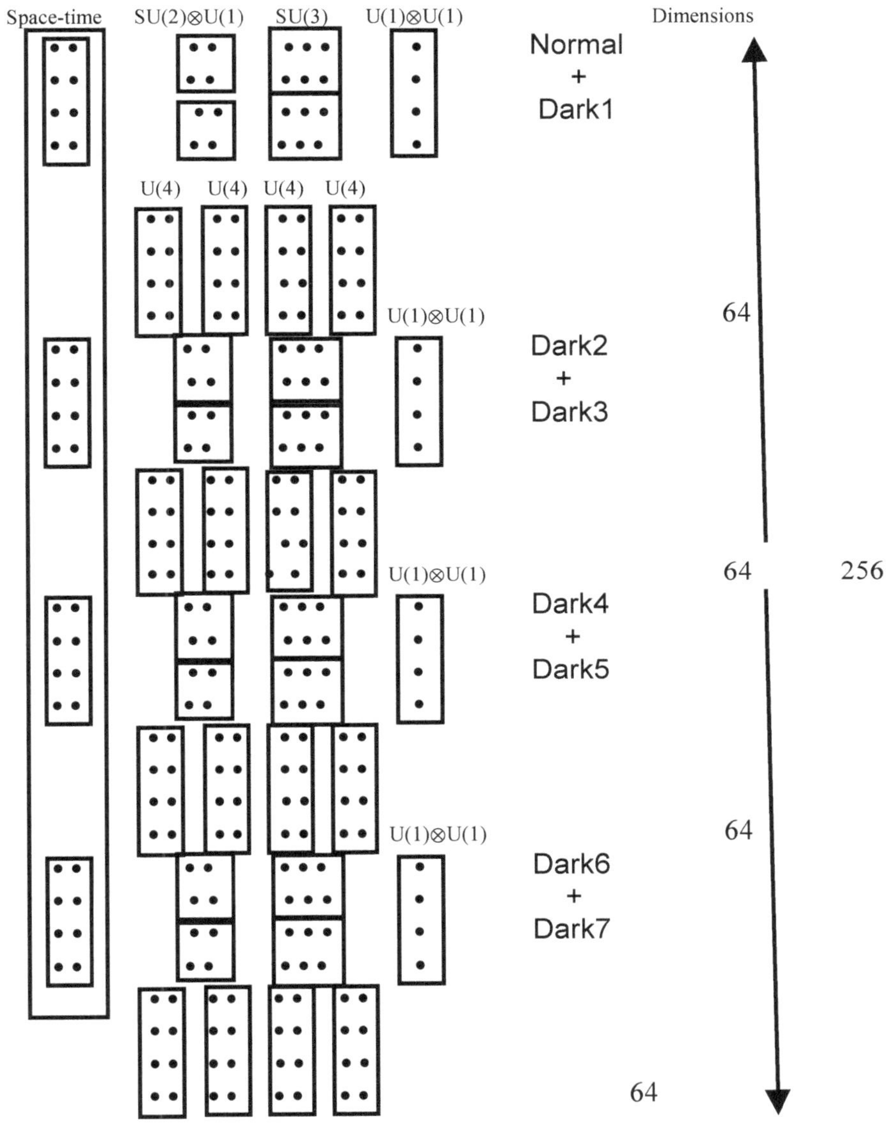

Figure 8.5 The 256 dimension groups of the four 8 $\times$ 8 blocks in each section of Fig. 8.4 Each 256 dimension section is one of the four layers of UTMOST dimensions.

9. The BMOST – UTMOST Connection

Universes are embedded within the Megaverse. A QUeST universe has four octonion space-time dimensions. The UTMOST Megaverse has eight complex octonion space-time dimensions. The BMOST space has 10 quaternion octonion space-time dimensions. The relevant Fig. 1.4 lines are:

4 Octonion Octonion (64) Maxiverse	Octonion Octonion	64×64	10 quaternion octonion
5 Quaternion Octonion[71] (32) Megaverses	Quaternion Octonion	32×32	8 complex octonion
6 Complex Octonion[72] (16) Universes	Complex Octonion	16×16	4 octonion

We can embed any number of QUeST universe instances within an UTMOST Megaverse.

We found we could define a basis for UTMOST, called BMOST, consisting of one seed urfermion residing in a ten dimension space-time (Chapter 7). A seed urfermion-antiurfermion pair annihilates into a Megaverse particle. This 10-space-time may be related to a yet deeper SuperString theory. Its details remain to be understood.

Thus a general foundation exists for EVERYTHING in the ten dimension Maxiverse.

---

[71] In our earlier books in 2020 we also designated this 1024 dimension space (5) as 64 complex octonion space.

[72] In our earlier books in 2020 we also designated this 256 dimension space(6) as 32 complex quaternion space.

10. Maxiverse Octonion Octonion Space: 10 Space-time Dimensions – With 32-Spinor urFermions

Octonion Octonion space is the ultimate source of Megaverses and Universes in our Octonion Cosmology. This space, that we call Maxiverse space, has 64 rows of 64 dimensions. Its dimensions total to $64^2 = 4{,}096$. Its entry in Fig. 1.4 is

4 Octonion Octonion (64)	Octonion Octonion	64×64	10 quaternion octonion

These dimensions are allocated to a ten quaternion octonion space-time (having 320 dimensions), and to fundamental representations of internal symmetry groups (occupying $4096 - 320 = 3776$ dimensions).

The internal symmetry groups can be viewed as consisting of four copies of the internal symmetries of the UTMOST Megaverse internal symmetries plus[73] $512 - 320 = 192$ dimensions allocated to additional internal symmetry groups. The 192 dimensions[74] can be divided into four sets of 48 dimensions and added to each of the four copies. The 48 dimensions in each modified copy give 12 dimensions to each layer of each copy.

Fig. 10.1 shows one layer of one copy of the UTMOST internal symmetry groups modified to include 12 additional dimensions. Octonion Octonion space has 16 copies of the internal symmetries in Fig. 10.1. The 12 dimensions can be initially allocated to a U(6), which may, in turn, be broken to the frequently found group U(1)⊗SU(2)⊗SU(3)⊗U(1) in QUeST and UTMOST (or perhaps SU(3)⊗SU(3)).

The Maxiverse dimension array can be partitioned into two hundred and fifty six $4 \times 4 = 16$ dimension blocks.

The physical role of this space is to support the derivation of Megaverse instances (particles) from urfermion-antiurfermion annihilation. The derivation is discussed in chapters 7 and 8. Urfermions are 10 dimension fermions.

---

[73] The quantity $512 = 4 \cdot 128$ is the total number of space-time dimensions in four copies of UTMOST. The number 320 is for 10 rows of 32 dimension (quaternion octonion) coordinates. The quantity $512 = 8 \cdot 16$ Megaverse space-time coordinates 4 Megaverses in Maxiverse

[74] There are 12 16-blocks in 192 dimensions. There are 3 64-blocks in 192 dimensions.

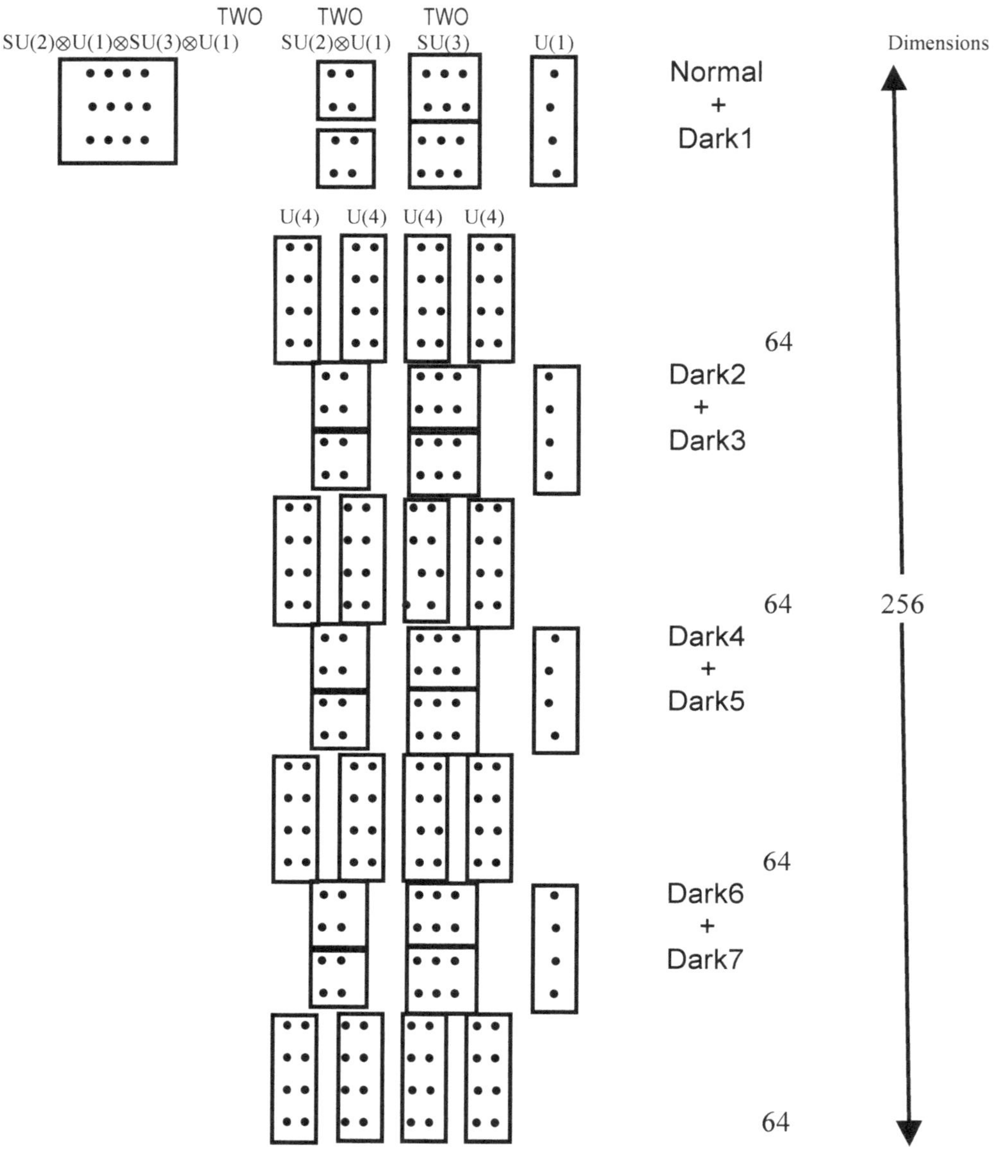

Figure 10.1. One layer of the four layers of one of the four copies of UTMOST dimensions *plus* an additional 12 dimension block with boxes around sets of dimensions for fundamental group representations. There are 16 copies of the internal symmetries listed here in Octonion Octonion space. Compare this figure to the UTMOST Fig. E.4 to see details of Normal and Dark sectors indicated above.

10.1 Block Structure of Universe, Megaverse and Maxiverse Spaces

The 16 dimension and 64 dimension blocks within octonion spaces is of some interest. In this section we provide block information for universe, Megaverse and Maxiverse spaces. As we saw earlier there are three types of 16 dimension blocks:

A. U(4)⊗U(4)
B. U(1)⊗SU(2)⊗U(1)⊗SU(2)
C. SU(3)⊗SU(3)⊗U(1)⊗U(1)⊗(Space-time).

These 16-blocks are repeated in the four layers of QUeST (Universe space) and the four layers of UTMOST (Megaverse space). They also are within a Maxiverse map.

There are also 64 dimension blocks (64-blocks) in QUeST and UTMOST and in the Maxiverse space as shown earlier. In QUeST each of the four layers is a 64-block.

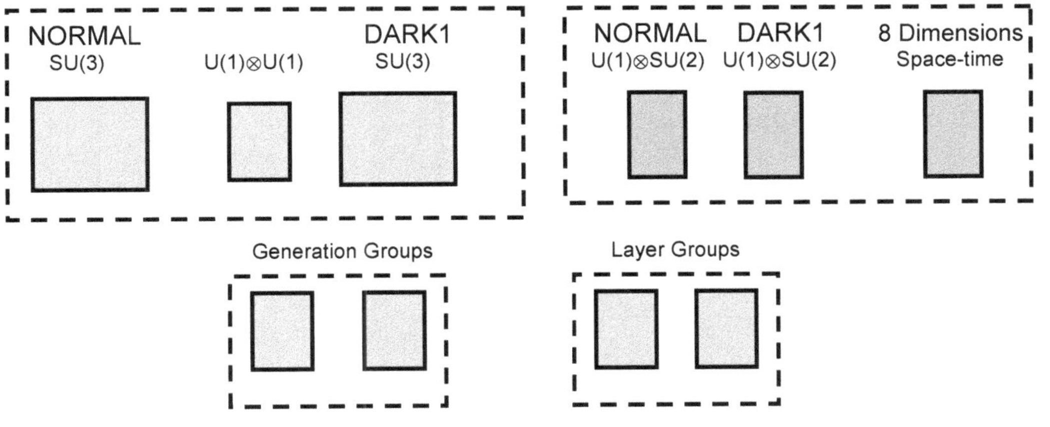

Figure 10.2. A 64 dimension (64-block) block. It is one layer in QUeST.

Space:		6 Complex Octonion Universe Space QUeST	5 Quaternion Octonion Megaverse Space UTMOST	4 Octonion Octonion Maxiverse Space
Dimensions		256	1024	4096
16-Blocks				
	Type A	8	32	128 + 6 = 134
	Type B	4	16	64 + 3 = 67
	Type C	4	16	64 + 3 = 67
64-Blocks		4	16	64 + 3 = 67
Space-Time Dimensions:		4 complex octonion	8 quaternion octonion	10 octornion octonion
Real Dimensions:		64	256	640

Figure 10.3. A table of the 16-blocks and 64-blocks in QUeST, UTMOST, and Maxiverse space. The additions for the Maxiverse result from using a 10 quaternion octonion space-time as shown sbove. The quantities above increase by factors of four except Maxiverse space where the quantities have additions due to the choice of space-time dimension.

Space:		3 Complex Octonion Octonion	2 Quaternion Octonion Octonion Space-timeless Spaces	1 Octonion Octonion Octonion
Dimensions		16,384	65,536	262,144
16-Blocks				
	Type A	512	2048	8,192
	Type B	256	1024	4,096
	Type C	256	1024	4,096
64-Blocks		256	1024	4,096

Figure 10.4. A table of the 16-blocks and 64-blocks of the three space-time-less spaces.

Space:		0 Superverse Space Complex Octonion Octonion Octonion
Dimensions		1,048,576
16-Blocks		
	Type A	32,768
	Type B	16,384
	Type C	16,384
64-Blocks		16,384

Figure 10.5. A table of the 16-blocks and 64-blocks of the Superverse space.

11. The Three Spaceless Spaces

The three formless, spaceless Spaces are the Octonion Octonion Octonion, the Quaternion Octonion Octonion, and the Complex Octonion Octonion spaces. Their entries in Fig. 1.4 are:

1	Octonion Octonion Octonion (512)	Octonion Octonion Octonion	512×512	0
2	Quaternion Octonion Octonion (256)	Quaternion Octonion Octonion	256×256	0
3	Complex Octonion Octonion (128)	Complex Octonion Octonion	128×128	0

They have $512^2 = 262{,}144$, $256^2 = 65{,}536$ and $128^2 = 16{,}384$ dimensions respectively. The spaces are defined to have no space-time part. They could be defined to furnish fundamental representations for U(131072), U(32768), and U(8192) respectively. Since they have no dynamics, being timeless and spaceless,[75] there can be no symmetry breaking. They have no elementary particles or interactions.

They can each be viewed as point spaces in the sense that they have no internal spatial or temporal distances.

They can also be viewed as exemplars for UST monads. We will consider monads in chapter 18.

11.1 Computation and a Three Spaceless Spaces Paradigm

Not infrequently physicists, and others, have suggested the universe is a great computer that runs by executing Physics equations. We now consider the possibility that the three spaceless spaces together with the other octonion spaces constitute a type of computer executing a program.

We define the computer and the executing dynamic program as having the general form of an executing C++ program based on all ten octonion spaces plus the Superverse space. We use the general form of an executing C++ program in computer memory.[76,77]

An executing C++ program in computer memory has several parts:

1. A part to define the classes and objects of the program. The corresponding part in the set of octonion spaces are spaces 4 through 10: Maxiverse space through the lowest Miniverse space. These spaces have the various symmetry groups and space-times defined within them (corresponding to

---

[75] This set of spaces can be viewed as an analogue of God after self–materialization. See Chapter 17.

[76] See S. Blaha, *C++ for Professional Programming* (International Thomson Publishing, Boston, 1995).

[77] A detailed example of the transformation of a quantum raising and lowering operator formalism to an Assembly Language formulation appears in *Cosmos and Consciousness* (Pingree Hill Publishing, Auburn, 1995 and 2002). A part on Quantum C Language has numerous interesting features such as a map between the address ponter & in C and C++, and a raising and lowering operator commutator expression.

C++ class definitions.) The spaces have Maxiverse, Megaverse, and Universe instances within them (corresponding to C++ objects.)

2. Another part of an executing C++ program in memory is the program code. We identify the program part of the Cosmos program as the contents of the spaceless spaces 1, 2, and 3. These spaces have instances can be assigned the terms and form of the physical equations governing the Cosmos. The instances embody some conceptual equivalent of the machine code for executing C++ programs.

We can *tentatively* assign programming roles to each of the three spaces.

Space 3

We will assume that we can assign a fermion, vector boson, or Higgs particle (letter,[78] label, variable) to each dimension. (In programming variables are often assigned numbers within an executing program.)

Some of the 16,384 dimensions of space 1 are assigned to the set of all particles (variables) in spaces 4 through 10 within the "Cosmos Program." The particle variables emerge from the fermions of the fundamental group representations, the vector bosons of the groups, and the Higgs bosons[79] of the fundamental group representations plus the graviton.

Space 1

Combinations of letters of space 3 can be used to define the complete set of interactions, as *words*, for all groups in spaces 4 through 10. The interactions of fermions, vector bosons and Higgs bosons are quadratic, tertiary, and quartic.[80] We note the set of interactions and interaction terms appearing in Physics equations is finite and much smaller than the 262,144 dimensions of space 1. .

Space 2

This space has two roles:

1. To connect the particles space 3 to the interactions of space 1.

2. To generate the dynamical equations of QUeST (UST), UTMOST, and of Maxiverse space and cause their execution through computation "steps" – an executing program counter.

---

[78] The discussions of Chapter 17 raise the possibility that complex dimensions can be viewed as analogues of letters (particles).

[79] Chapter 16.

[80] Gravitons must be treated separately.

11.2 Analogue to the *Sefer Yetzirah* and the Logical God-View in Chapter 17

Chapter 17 describes the three aspects of God in the *Sefer Yetzirah* (as well as later schools of thought.) An analogue to the above computation view is apparent.

Chapter 17 also describes part of the author's God Theory[81] which also displays a threefold view of God based on Logic including the such concepts as God's Omnipotence, Predestination and Free Will.

[81] Blaha (2018e).

12. The Minispaces

Instances of these four spaces are generated by the annihilation of fermion-antifermion pairs. All these spaces have 16 dimensions. Their entries in Fig. 1.4 are:

7 Quaternion (4)	Quaternion	4×4	4 Real
8 Real (4)	Real (4)	4×4	4 Real
9 Real (4)	Real (4)	4×4	4 Real
10 Real (4)	Real (4)	4×4	0

QUeST fermions and antifermions have 4-spinors that are used to construct the 4×4 dimension array for Minispace 7. Minispace 7 has four real coordinates and 4-spinors for its fermions. The internal symmetry can be specified as U(1)⊗SU(2)⊗SU(3)⊗U(1) where the last U(1) factor is a Fermion symmetry group.

The next two Minispaces, 8 and 9, are the same as space 7. Their instances are generated by a series of fermion-antifermion annihilations.

Space 10 is assigned no space-time by construction to preclude an infinite sequence of further nested minispaces. It has no fermions or bosons. It has a subset of QUeST internal symmetries. It does not experience dynamical evolution. Consequently there are no interactions or symmetry breaking.

A minispace instance has a location and momentum in a space-time instance of the space above it.

An instance of space 7 could be detected experimentally as an annihilation event with no final state particle produced that ultimately decays into a fermion-antifermion pair. Or an experiment may see the creation of a fermion-antifermion pair from "nothing." No evidence for these phenomena has been found to the author's knowledge.

13. The Superverse Space

The Superverse is an all-encompassing space that can have the 10 lower spaces of Fig. 1.4 as subspaces. It has no intrinsic space-time or internal symmetries although subspaces can have instances with space-times and internal symmetries. Its entry in Fig. 1.4 is:

0 Complex Octonion Octonion Octonion (1024)	Complex Octonion Octonion Octonion	1024×1024	0

The Superverse has $1024 \times 1024 = 1{,}048{,}576$ dimensions. It fits naturally within the pattern of spaces in Fig. 1.4. Spaces 1 through 10 can be viewed as subspaces of Superverse space. See Fig. 1.6.[82]

[82] In chapter 17 we consider the possibility that Superverse space is the analogue of God (Ein Sof) before self–materialization when God is in a formless state of Atzmus (Devine Essence) (Hasidic Wisdom). Fig. 1.6 then reflects the withdrawal to create an empty space.

14. Strings in the Maxiverse

The Maxiverse (space 4), having 10 space-time dimensions, supports strings. These strings can be manifest in a Megaverse and/or universe upon restriction of the number of dimensions of the string. Strings within space instances 5 through 10 may be generated by fermion-antifermion annihilation as discussed earlier.

15. Bags and Miniverses

The Octonion Cosmology spaces 4 through 10 supports a variety of constructs:

1. A universe can have Miniverse instances (bubbles), which may be “bags” of the type envisioned by the MIT bag model.

2. A Miniverse instance of any of spaces 7, 8, 9 may contain instances of “lower” spaces, also forming types of bags.

The space instances within “smaller” instances can be generated by fermion-antifermion annihilation as discussed earlier.

16. Higgs Boson Sector of Octonion Cosmology

The Higgs bosons of The Standard Model and the Unified SuperStandard Theory (UST) also appear in QUeST and in UTMOST. We saw that these octonion theories naturally had the symmetry groups of UST. The fundamental representations of the groups defined the symmetries and their sets of gauge vector bosons. They also defined the fermions associated with the fundamental group representations. The resulting number of fundamental fermions equaled the total number of internal symmetry dimensions of space 6 for QUeST (Universe space) and of space 5 for UTMOST (Megaverse space).

16.1 The Number of Higgs Bosons in QUeST and UTMOST

In this chapter we show that each fundamental internal symmetry representation may have a set of Higgs bosons. Thus the total number of Higgs bosons equals the total number of internal symmetry dimensions of space 6 for QUeST and of space 5 for UTMOST. *For both spaces, the number of Higgs particles generated in this manner equals the number of fundamental fermions associated with the space.*

16.2 Higgs and Symmetry Breaking

The sets of Higgs bosons are divided into two parts: those that embody symmetry breaking, and those that do not. The set of Higgs bosons that do not embody symmetry breaking consist of the Higgs bosons of SU(3) internal symmetry groups since SU(3) gauge vector bosons are massless. These Higgs fields may, however, give mass to SU(3) quarks.

The set of Higgs bosons that generate symmetry breaking consist of the Higgs bosons of the other (non-SU(3)) internal symmetry groups. The Higgs bosons in this set give masses to the vector bosons of these groups. In addition, these Higgs bosons give masses to the corresponding fundamental fermions.

17. Octonion Cosmology Parallels in *Sefer Yetzirah* Cosmology, Linguistics, and God-View

Creation mirrors the Creator.

Having explored the octonion spaces, we now venture into the transcendental by considering parallels of Octonion Cosmology to the *Sefer Yetzirah*. The *Sefer Yetzirah* has been described by some as Metaphysical, and by others as mathematical and linguistic. We will see parallels in it to Cosmology, Linguistics, and a view of God.

17.1 An Exegesis of the *Sefer Yetzirah*

The revered *Sefer Yetzirah* (*Book of Formation*) of Judaism contains a cosmology that corresponds in part to the Octonion Cosmology of this author. It is viewed by many as the work of the Patriarch Abraham (Second Millennium BCE) due to divine revelation from his Friend, God.[83] Others view it as redacted to its present form by Rabbi Akiva. Akiva's version dates back to the Second Century BCE according to some; giving a date of 100 – 300 CE.

In analyzing the *Sefer Yetzirah* for its physical-cosmological content we must map corresponding physical and qualitative features bearing in mind the state of physical discussion 2000 years ago.[84]

17.2 Physical Concepts in the *Sefer Yetzirah*

The *Sefer Yetzirah* has many concepts about the formation of the Cosmos. We find:

A series of emanations of mediums (planes of existence) between god and the universe

Thirty-two mysterious paths of wisdom emanations of the Sefirot

Ein Sof – infinity formless state unchanging

Thirty-two mysterious paths of wisdom

Ten Sefirot without limits (text: "ten and not nine, ten and not eleven"[85] suggesting that the number 10 is of special significance and is to be remembered

---

[83] Jewish lore attributes it to Adam, then to Noah, and then to Abraham.

[84] This sort of map is also of significance in the history of Greek Science and Philosophy in PreSocratic times. This author pointed out that "Science" based on the actions of Gods was mapped (transformed) to a more physical Science based on types of matter in PreSocratic times. Blaha (1964) unpublished.

[85] *Sefer Yetzirah.*

in future), "ten calibrations of empty space"[85] suggesting 10 space-time coordinates.

Gematria - Sefirot have numerical values – they are calibrations

Sefirot Twenty-two basal letters - 22 channels

Sefirot emanate from above to below

First Sefirot is Keter. It has 3 levels within

3 mothers and fathers

The Kabbalah extends the set of *Sefer Yetzirah* concepts:

Four worlds or planes of existence:

- Emanation
- Creation
- Formation
- Action

Intermediate forms (worlds) from the beginning to our universe
Forms are analogues to the Creator
Key numbers are 3 7 12
A correspondence between spiritual and physical

17.3 Cosmology of *Sefer Yetzirah*

The *Sefer Yetzirah* postulates a beginning with 10 Sefirot (without limits), which generate a set of 32 emanations (mysterious paths of wisdom).

There are four mediums (four worlds or planes of existence), which extend from God to the universe.

God is the Ein Sof indivisible yet composed of parts: the Emanator of the 10 Sefirot: Unending, Unchanging, infinity.

The first three of the Sefirot (Ayin, Ein Sof, Ohr Ein Sof) constitute the Godhead.

Summary of Formation

Godhead
- One God

Composed of three elements
- Unending, Unchanging, infinite
- Emanator of the Ten Sefirot plus 22 letters
- Created His world (universe) by 3 derivatives

Four Worlds
- Ten Sefirot without limits - enumerations
- 32 Emanations that extend from above to below

22 channels
One world is the universe

17.4 Comparison of Its Cosmology with Octonion Cosmology

The *Sefer Yetzirah* has a Cosmology as seen above. The "translation" of that Cosmology into a physical theory requires a map from its qualitative-religious vocabulary to physical concepts.

As we noted earlier there is a correspondence to the Octonion Cosmology of 10 spaces. Comparing to Fig. 1.4 (and subsequent chapters) we see a correspondence:

Ein Sof	↔ 1024×1024 Superverse
Three elements of the Godhead	↔ Three spaces 1, 2, and 3; unending, unchanging, infinite, without physical content
Ten Sefirot	↔ The 10 octonion spaces. Also the 10 space-time dimensions of space 4 Enumerations
Thirty-two emanations	↔ the 32×32 dimension array space 5 A Megaverse space
Twenty-two Channels	↔ the part of space 5 for internal symmetries that extends below to spaces 6 – 10; 22 = 32 – 10 Channels
First two derivatives (1st and 2nd worlds)	↔ spaces 4 and 5
The third derivative ((3rd world)	↔ universe space 6
The 4th world	↔ spaces 7, 8, 9, 10 subspaces of space 6

Figure 17.1. Map between *Sefer Yetzirah* and octonion spaces of Fig. 1.4.

Fig. 17.2 contains the features of the 10 spaces reproduced from Fig. 1.4 for the reader's convenience. Fig. 17.3 presents a graphic depiction of the 10 spaces based on the map of Fig. 17.1.

17.5 Syncretic Formulation of *Sefer Yetzirah* and Octonion Cosmology

The process of actualization of the parts of the 10 spaces must be viewed as instantaneous since a time does not exist outside the spaces.

0. The process begins with the .self–materialization of God (Ein Sof) from a formless state of Atzmus (Devine Essence) (Hasidic Wisdom) by a

withdrawal to create an empty space (according to Lurianic Kabbalah). God corresponds to the 1024 × 1024 Superverse before self-materialization. The Superverse is infinite, unchanging and formless since there has no dynamics. The withdrawal corresponds to the breakdown of the 1024 × 1024 Superverse into the 10 octonion spaces as shown in Fig. 1.3.

1. The 10 Sefirot are self-materialized from God. They can be viewed as analogues of the 10 octonion spaces. They are calibrations.

2. The three elements of the One God (after self-materialization) correspond to the top three octonion spaces.

3. The 4th space, Maxiverse space, has 10 space-time dimensions and 32 emanations (32-spinors) for the fermion particles within it. Fermion-antifermion annihilation generates instances of space 5.

4. Space 5 instances are Megaverses (Multiverses). Fermion-antifermion annihilation in space 5 instances generates instances of space 6 – universe space instances.

5. The 22 emanations from space 4 generate the internal symmetry groups of spaces 5 – 10. The other 10 emanations become 10 space-time dimensions.

6. Spaces 4, 5, and 6 are three of the "four worlds." The 4th world consists of the union of the subspaces 7, 8, 9, and 10.

7. The Cosmos is formed! See Fig. 17.3.

Spectrum Number

Coordinate Type	Number of Coordinates	Dimension Array Size	Space-Time Dimensions
Superverse			
0 Complex Octonion Octonion Octonion (1024)	Complex Octonion Octonion Octonion	1024×1024	0
Spaceless[86]			
1 Octonion Octonion Octonion (512)	Octonion Octonion Octonion	512×512	0
2 Quaternion Octonion Octonion (256)	Quaternion Octonion Octonion	256×256	0
3 Complex Octonion Octonion (128)	Complex Octonion Octonion	128×128	0
Cosmology[87]			
4 Octonion Octonion (64) Maxiverse	Octonion Octonion	64×64	0 quaternion octonion
5 Quaternion Octonion[88] (32) Megaverses	Quaternion Octonion	32×32	8 complex octonion
6 Complex Octonion[89] (16) Universes	Complex Octonion	16×16	4 octonion
Minispaces[90]			
7 Quaternion (4)	Quaternion	4×4	4 Real
8 Real (4)	Real (4)	4×4	4 Real
9 Real (4)	Real (4)	4×4	4 Real
10 Real (4)	Real (4)	4×4	0

Figure 17.2. The Superverse and the spectrum of ten octonion spaces shown in Fig. 1.4. In total they are 11 spaces and correspond to the 11 Sefirot formulation of later schools.

17.5.1 Sefirot Groupings

The Sefirot are often grouped into three parts plus one. Similarly, the ten spaces of Octonion Cosmology may also be grouped in that manner. The first group consists of the spaceless spaces 1, 2, 3. The second group consists of the Maxiverse, Megaverse, Universe and three Miniverse spaces 4, 5, 6, 7, 8, 9. The third group is the Miniverse space 10.

---

[86] Spaceless spaces are spaces without a space-time.

[87] Cosmological spaces are those, which are directly related to physics from megaverses and universes to elementary particles.

[88] In our earlier books in 2020 we also designated this 1024 dimension space as 64 complex octonion space.

[89] In our earlier books in 2020 we also designated this 256 dimension space as 32 complex quaternion space.

[90] A Minispace is a subspace of a universe space.

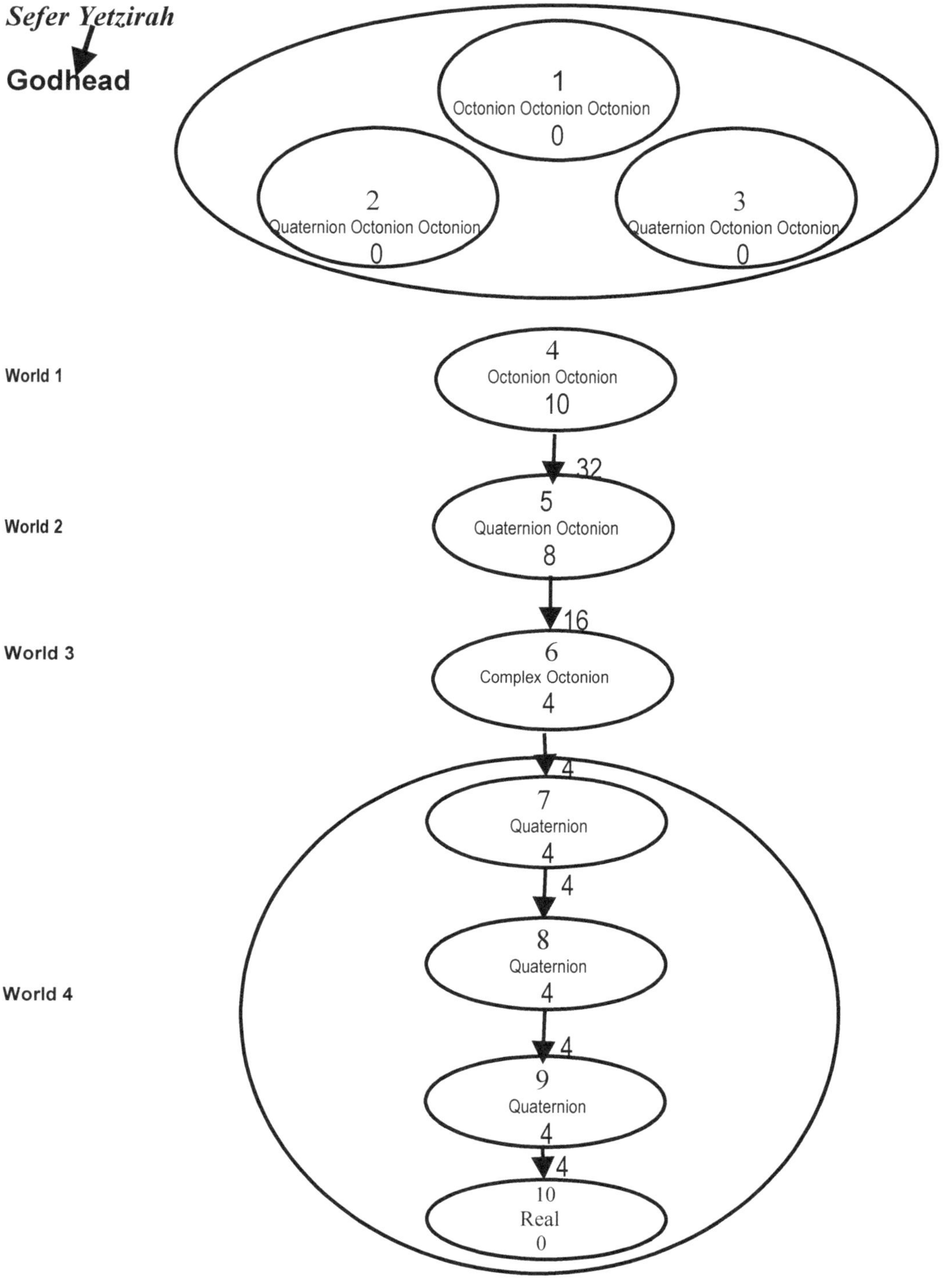

Figure 17.3. The descent to below of the ten spaces with their spectrum number and their space-time dimensions indicated within each oval. The set of 10

spaces reside within a space of 1024 × 1024 dimensions (the Superverse) that is not displayed.

17.6 Origin of *Sefer Yetzirah* Cosmology

The *Sefer Yetzirah* dates back 2000 – 4000 years according to lore. Its cosmological content is analogous to Octonion Cosmology if translated into physical terms and suitably interpreted as above.

The Formation point of the *Sefer Yetzirah* appears to be the first instant of the first day of the Seven Days of Creation in the Torah (Bible). Time begins at that point. Time does not exist before that point. Thus the *Sefer Yetzirah* complements the Biblical account of Creation.

The great question is to understand the origin of the *Sefer Yetzirah*'s concepts. Clearly they are based on linguistics in part as well as analogies to common things of experience.

Yet the combination of its concepts, and numerics, bespeaks information from external source(s). The *Sefer Yetzirah* is attributed to the Patriarch Abraham (Devine Revelation) who received it, according to lore, from his Friend, God. However the possibility of an origin from other sources is not excluded. The close similarity of Octonion Cosmology to the Cosmology of the *Sefer Yetzirah* is striking!

17.7 Parallels in Linguistics with Octonion Cosmology

The *Sefer Yetzirah* also presents a Linguistics Theory based on the 32 Emanations. The 10 Sefirot and the 22 channels are mapped to letters of the Hebrew alphabet.

Remarkably the fundamental representations of groups of a layer in QUeST also number 32 complex dimensions (Fig. 17.4).

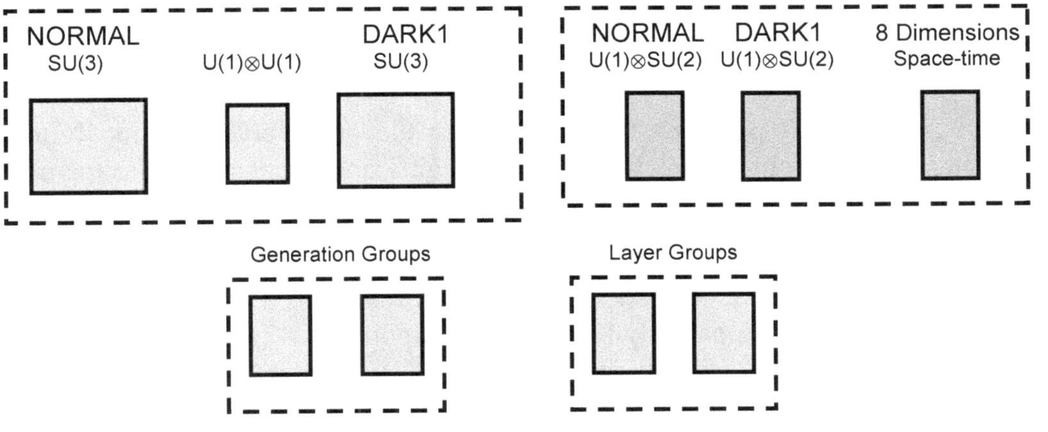

Figure 17.4. A 64 dimension 64-block has 32 complex dimensions. It is one layer in QUeST. The 32 = 10 Sefirot plus 22 channels correspond to 32 complex dimensions (64 real-valued dimensions) of a QUeST layer.

The correspondence between letters and complex dimensions is clear.

Based on this correspondence we can create a Octonion Cosmology analogue to the 32 emanations (Fig. 17.5). The 10 Sefirot analogue contains Fermion and Layer groups. These groups are "higher" in the sense that they pertain to layers in QUeST. The groups in the analogue of the 22 channels are "lower" in the sense that they relate to aspects within a given layer in QUeST.

The clean breakup of the groups of the 22 channels into 3, 7, and 12 parts as displayed in Fig. 17.5 corresponds to the divisions in the Hebrew alphabet. Again we find a remarkable correspondence between complex dimensions and letters.

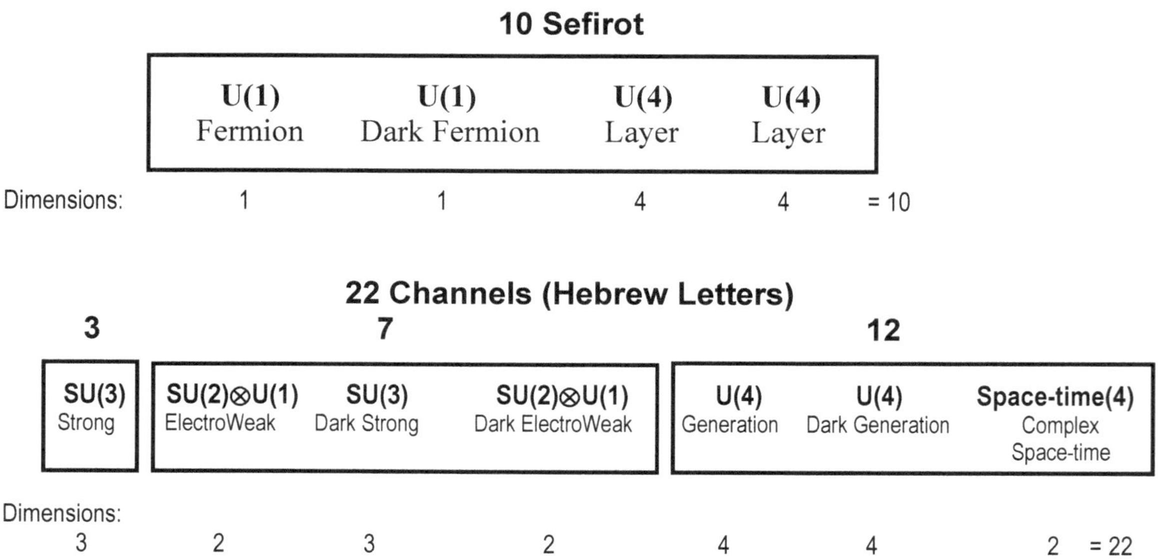

Figure 17.5. The 32 emanations—10 Sefirot and 22 channels (letters)—related to the fundamental representations groups of a QUeST layer.

The *Sefer Yetzirah* text passage: "twelve and not eleven, twelve and not thirteen"[85] strongly suggests the number 12 is of special significance and must be remembered. One wonders if it is perhaps a message to future generations.

17.7.1 The 231 Gates in the Wall

The Linguistics passage of the *Sefer Yetzirah* refers to "a wall with two hundred and thirty one gates."[85] It apparently refers to the combinations of 22 channels (Hebrew letters). There are $22{\cdot}21/2 = 231$ combinations of two letters ignoring order. Each combination is a gate. The analogue to a "gate" in Octonion Cosmology is a two complex dimension pair. It is natural to regard the two complex dimension pair as the four dimensions of 4D space-time. Thus the 231 enumerations[91] of two complex dimensions form the set of all possible space-time sets within an Octonion Cosmology

[91] A gate therefore is an aperture to space-time.

QUeST layer that are obtained by rearrangement of the 22 complex dimensions. Fig. 17.5 shows the space-time part of the 22 channels among the 12 channels.

The 231 gates enumerate all possible sets of four dimensions (2 complex dimensions) in the 22 channels of a layer. We again find a remarkable correspondence with the *Sefer Yetzirah*!

17.8 Parallel of *Sefer Yetzirot,* and Later Schools, God-View to Octonion Cosmology

Octonion Cosmology has parallels to the God-View of the *Sefer Yetzirot,* and Later Schools. The parallels are:

God	Octonion Cosmology
Formless state of Atzmus (Devine Essence) (Hasidic Wisdom) Endless, formless, no attributes, nothingness, nameless	Superverse Space No space-time, no symmetry groups Formless
Withdrawal to create an empty space (according to Lurianic Kabbalah)	"Generation" of 10 Octonion Spaces 1 through 10; Diagonal blocks within the Superverse space
10 Sefirot	The 10 Octonion Spaces
Three Sefirot: Ayin, Ein Sof, Ohr Ein Sof	Octonion s[paces 1, 2, 3 No space-time, no symmetry groups Formless

The closeness of the parallels is clear!

17.9 Parallels to *Sefer Yetzirah* Cosmology, Linguistic Theory & God-View

There are clear parallels to aspects of Octonion Cosmology and Physics in the preceding discussions.

17.10 Sefer Yetzirah Hints of Lost Ancient Physics

The preceding discussions in this chapter clearly are indications of a lost knowledge of Physics (and Mathematics). The origin of this Physics is unclear if it is not according to the Lore of an origin with Patriarch Abraham as perhaps modified by Rabbi Akiva.

17.10.1 The 231 Gates to Space-Time

A particularly important point is the 231 gates in the wall that indicate knowledge of four-dimensional space-time, and also indicate that any two complex dimensions in universe space (space 6) are suitable for space-time.

17.11 Cascading Quantum Probability Tweaks to Manage History Assuming God

The introduction of chance in physical events opens a "new" possibility for God's intervention in Cosmic events.[92] Without chance God's intervention would be strictly limited by Physical Law (which God established absolutely) making "miracles" very limited, and in most cases impossible.

With chance in physical events, it becomes possible for God to "tweak" an event quantum mechanically[93] to make a certain result happen. This result can then cause a series of further God-manipulated quantum events to happen with perhaps additional Quantum tweaks and so on.

The result would be a cascading set of events caused by Quantum tweaks consistent with Quantum probability, and Physical Law—a chain reaction of events that can generate a miracle or even change history. One remembers here the lament of Richard III of England, "My kingdom for a horse." (We will not comment on the chain of events that would have put a horse within Richard's grasp.)

17.11.1 Quantum Source of Free Will

Free will is usually viewed from the aspect of a person's mental state. However, free will has several forms. Some forms relate to the person's perceived situation with free will exercised to maximize the person's benefit and/or views. This form involves a significant part of the person's consciousness.

Another form of free will is chance decisions. A good example of this form is spot decisions as to where to place chips on a roulette table. This type of decision can be based on guessing and may reflect a chance event in a person's mind. Free will based on chance miniscule events in a person's mind can be the result of cascading Quantum tweaks.

Thus a person can have free will without God's intervention. But God can also intervene by making Quantum choices of outcomes in a person's neurons.

God can thus intervene in the Cosmos to implement change, guide free will, and execute miracles through Quantum effects.

17.12 A Logical View of God

The author has recently developed a God-View based on Logic and Physics in Blaha (2018e) entitled *Unification of God Theory and Unified SuperStandard Model THIRD EDITION*. This view takes the generally accepted features of God and develops a God-View of a God with a composite nature. The God-View is consistent with that of section 17.8. It explains the threefold nature of God. This God-View leads to an understanding of Free Will and Predestination within a Quantum Framework.

The remainder of this chapter is an excerpt from Blaha (2018e).

G-6.1 God and the Cosmos

In earlier chapters we saw that God (the Unmoved Mover) requires the Cosmos to have a dynamic aspect. The Unmoved Mover specifies the nature of

[92] An extract from section G-A.2 below.

[93] God selects one of the possible probabilistic outcomes to suit God's purpose.

the Cosmos through the considerations of chapter G-4 which lead to the prerequisites and axioms presented in Chapter 0. The derivation of The Theory of Everything in the second section of this book (and Blaha (2018a) and (2018b)) yields the Standard Theory of Elementary Particles plus much more. The Unmoved Mover determines and causes the dynamics (the "Movement") of the evolving Cosmos. As we saw earlier the Cosmos exists forever so the Creation of the Cosmos by the Unmoved Mover, although it is done by God, is "before time" since time is an artifact of the Cosmos. Time did not exist before Creation. Thus Creation does not happen at some instant but really exists forever as an "implicit" part of God (with the Unmoved Mover).

G-6.2 Is the Cosmos Part of God?

Or does God permeate the Cosmos? If the Cosmos is a distinct part of a Godhead composed of the Unmoved Mover, the Cosmos, and perhaps additional parts then we must ask how the Cosmos partakes of God-like features. If we regard the Cosmos as merely matter and energy, as Physicists view it, then there is no Godliness in the Cosmos. So an enhanced theory of the Cosmos would be required, such as that of Leibniz,[94] who suggested particles of matter and energy (monads) have a divine "spark." An enhanced theory of the Cosmos would put the Cosmos on a God-like footing and unite It with the Unmoved Mover. See Table G-6.1 (below) for a comparison of features of God and the Cosmos. The fact that they both exist forever is a strong argument for a Cosmos that is part of God. Spinoza suggested a similar pantheistic view of the Cosmos. Karl Krause coined a name for a version of this view, which he called *panentheism*.

A panentheistic view of the Cosmos distinguishes the Cosmos from the Unmoved Mover. So we must consider the possibility that the Godhead has at least two parts: the Unmoved Mover and the "panentheistic" Cosmos that acquires "Godhood" in its particles of matter and energy—perhaps through divine sparks (monads?) or a similar mechanism.

The godliness of the Godhead then is due to the godliness of Its parts. A primary feature of godliness is its infinity of existence. The Unmoved Mover and the Cosmos both have an infinite existence.

G-6.3 God's Relation to the Cosmos – The Connection

We have distinguished the Unmoved Mover from the Cosmos with both partaking of the Godhead. The Unmoved Mover is connected to the Cosmos by causing natural laws to be obeyed throughout the Cosmos. It is also possible that the Unmoved Mover, by exerting Its will and unlimited power, may cause events (interventions) to take place within the Cosmos that we might view as miracles or at least strange phenomena. (Appendices G-A and G-G describe a mechanism for God to act in the Cosmos to achieve Its will without violating the Physical laws of the Unified SuperStandard Theory.)

---

[94] Leibniz viewed all particles of matter as containing a devine spark – his *monad* theory. See Rescher (1967).

The *Connection* between the Unmoved Mover and the Cosmos could be viewed as simply a "transmission" between them. It might also be viewed as an entity with God-like properties. From previous chapters we see the Connection must be eternal—without beginning or end. The Connection must partake of the divinity of the Unmoved Mover and the Cosmos to achieve dynamical control of the Cosmos by the Unmoved Mover, and to support intervention by the Unmoved Mover in the Cosmos. The method by which it achieves these goals is in Its interaction between the Unmoved Mover and the Cosmos.

If the Cosmos has sparks of divinity, then the Connection—connecting both divine parts—must have a divine aspect. Its "infinitesimal" parts must also have a spark of the divine.[95]

Thus we arrive at *a triune Godhead consisting of the Unmoved Mover, the Cosmos, and the Connection between them*. Table G-6.1 compares the features of the parts.[96] Fig. G-6.1 diagrams the interrelations of the parts in the Godhead.

Is there one God? Yes, according to our assumption and ensuing discussion. We merely provide a logical separation of God into parts.

In Appendix G-G we discuss the nature of the Connection in some detail within the framework of our Unified SuperStandard Theory. Lay readers may choose to skip this somewhat technical chapter since it does not play a role in the discussions in succeeding chapters.

---

[95] Perhaps as Leibniz-like monads. See Appendix G-G for a "particulate" form of Connection. The particles of the Connection would have a divine spark just as the particles of the Cosmos.

[96] One cannot say that the three parts of the Godhead are as separated as our analytical discussion suggests. They may well be inseparable. Also, the possibility of other parts is not precluded.

ENTITY	NATURE	LIFETIME	ROLE
Unmoved Mover	Spirit	Forever	Creates and Causes Dynamics
Cosmos	Spirit/Material	Forever	Scene of dynamic Activity
Connection	Spirit	Forever	Connects God, as Unmoved Mover, to the Cosmos, and intervenes in Cosmos events

Table G-6.1. The entities of existence. The Cosmos could have been viewed as a spirit if all forces between its parts are zero. In reality there are forces and so the Cosmos is "material" as well.

Our major findings are:

1) The part of God called the Unmoved Mover needs the Cosmos.

2) The Cosmos exists forever in the past and the future. It is divine just as the Unmoved Mover. Together they are combined in God.

3) The Unmoved Mover and the Cosmos are 'joined" by a part called the Connection which exists forever and is divine as well. The Connection is part of God.

4) The Connection implements the dynamics of Physical Laws in the Cosmos and performs any intervention required by the Unmoved Mover.

5) The Godhead consists of at least three divine parts: the Unmoved Mover, the Cosmos, and the Connection between them. Together the parts functionally form a triune God.

6) The Cosmos and Connection parts of God have a God-like nature that is represented in the Cosmos within the dynamics of the Unified SuperStandard Theory or a similar Theory of Everything.

7) The Godliness present in the Cosmos, and the Connection, resides in a divine "spark" within the particles[97] of matter and energy within them—a form of panentheism.

8) The apparent contradiction between Predestination and Free Will is resolved by Quantum Theory.

Appendix G-A. Predestination, Free Will and Quantum Theory

Fixed Fate, free will, foreknowledge absolute.
J. Milton, *Paradise Lost*, Book ii, line 560

Predestination is a troubling concept that is accepted in some religions and rejected by others. It states that all events happen through the design of God and that free will does not truly exist—free will is viewed as an illusion.

In this chapter we will suggest that Quantum Theory enables free will to exist, enables God to intervene in the Cosmos without violating Physical Laws, and yet enables God to be all-knowing of events past and future in all detail with "foreknowledge absolute" through the Tape of Fate.

G-A.1 God's Tape of Fate of the Cosmos and Predestination

Earlier we suggested the view that God, in effect, has an infinite tape containing all aspects of the Cosmos from the remote past to the remote future. His foreknowledge indicates that he is fully aware of the entire contents of the tape everlastingly in the past and future.[98]

God's "foreknowledge absolute" is the predestination to which many religions refer. It appears to eliminate the possibility of arbitrary intervention by God into Cosmic events at selected points in a manner not present in the tape. Yet God's intervention does occur according to all religious thought and any intervention by God must appear in the Tape of Fate since it is a faithful record of eternity.

God cannot arbitrarily and randomly intervene in Cosmic events because God's Physical Laws[99] cannot be violated—superficially contradicting God's assumed limitless power. A moment's thought shows that all of God's

[97] The divine spark resides within the Qubes and Qubas in particles and energy in our Unified SuperStandard Theory.
[98] Time travel, which appears to be possible in principle, can be accommodated within the Tape of Fate as simply a flattening of the tape segments for a time travel trip to one composite tape segment—a multiplexed tape segment.
[99] God's laws are created and required by the Unmoved Mover. They cannot be violated. (All Physical experience confirms this point.) They were created by God! But Quantum effects, because they are probabilistic in nature, can be selectively used by God to achieve God's "goals."

interventions at various times in the "infinite life" of the Cosmos must have existed as a reality forever—from the infinite past to the infinite future of the Cosmos. Thus we have the issue of intervention without violation of Physical Laws.

G-A.2 The Need for a Quantum Cosmos

In Blaha (2018a) we suggested that Quantum Theory was the simplest way to ensure that matter and energy were both made of discrete particles.

We now consider another, perhaps deeper, justification for Quantum Theory. While Quantum Theory has many facets, one facet of great importance in our view, is that it introduces chance into physical processes. When an event takes place, such as the collision of two particles, there can be varying results of the collision. Each possible result has a certain probability of happening—just as flipping a coin has a probability of "coming up" heads and a probability of "coming up" tails.

G-A.2.1 God's Intervention in the Cosmos

The introduction of chance in physical events opens a "new" possibility for God's intervention in Cosmic events. Without chance God's intervention would be strictly limited by Physical Law (which God established absolutely) making "miracles" very limited, and in most cases impossible.

With chance in physical events, it becomes possible for God to "tweak" an event quantum mechanically[100] to make a certain result happen. This result can then cause a series of further God-manipulated quantum events to happen with additional Quantum tweaks and so on.

The result is a cascading set of events caused by Quantum tweaks consistent with Quantum probability, and Physical Law—a chain reaction of events that can generate a miracle or even change history. One remembers here the lament of Richard III of England, "My kingdom for a horse." (We will not comment on the chain of events that would have put a horse within Richard's grasp.)

G-A.2.2 Quantum Source of Free Will

Free will is usually viewed from the aspect of a person's mental state. However, free will has several forms. Some forms relate to the person's perceived situation with free will exercised to maximize the person's benefit and/or views. This form involves a significant part of the person's consciousness.

Another form of free will is chance decisions. A good example of this form is spot decisions as to where to place chips on a roulette table. This type of decision can be based on guessing and may reflect a chance event in a person's mind. Free will based on chance miniscule events in a person's mind can be the result of cascading Quantum tweaks.

---

[100] God selects one of the possible probabilistic outcomes to suit God's purpose.

Thus a person can have free will without God's intervention. But God can also intervene by making Quantum choices of outcomes in a person's neurons.

God can thus intervene in the Cosmos to implement change, guide free will, and execute miracles through Quantum effects.

On occasion God loads the dice.

But as Einstein suggested, "Not maliciously."

G-A.3 Quantum Observer Problem

Having brought in Quantum Theory to support intervention by God we must now consider an important Quantum Theory problem that is resolved by the existence of God.

The problem is the need for a Quantum Observer to view an event and to observe which of the possible results actually occurs in reality. A famous example is Schrödinger's Cat. A cat is placed in a box with a radioactive element that decays after a "random" time. The setup in the box causes the cat to die when the radioactive element decays. The problem is: After x minutes is the cat alive or dead? The answer can only be found if the box is opened by an observer. Before that point in time the cat is in an unknown state with the reality of its life in doubt.

Thus reality becomes an issue of observation. Some have suggested that complete knowledge of the entire element-cat-box apparatus as a quantum state would resolve the issue. But that is unlikely because the composite system is quantum and subject to chance as well.

One could extend the apparatus state to the entire universe. But that would still be quantum.

So the tenure of the cat's lifetime remains uncertain forever. A sad situation.

The situation of Schrödinger's Cat applies in processes throughout the Cosmos. As a result it appears that reality is an ever present issue for the Cosmos.

G-A.4 God as the Ultimate Quantum Observer

The only clear-cut solution to the reality problem is God—the Ultimate Quantum Observer.[101] God need only look at "the Tape of Fate" to see the results of all Quantum events. Thus God gives reality to the Cosmos. An infinite progression of observers is thus avoided.

---

[101] Recently generalizations of the Schrödinger's Cat "experiment" have been considered in which an observer is placed in the box. Then it appears that if two observers, outside the box, open the box they will find different observations making a quantum paradox. This controversial result has been the source of much discussion..Again, our choice of God as the Ultimate Quantum Observer eliminates the paradox. See D. Frauchiger and R. Renner, arXiv:1710.05033v1 (2017) and arXiv:1604.07422v1 (2016) for detailed discussions of the paradox.

G-A.5 Events Without Observers

A question raised in some religions: does an acorn make a sound when it falls to the ground if no one is present to hear the sound, is also resolved. God, the Unmoved Mover, is the omniscient Observer in all quantum and non-quantum events. It makes reality real whether or not a human observer is present.

G-A.6 God Knows All at All Times

All eternity is predestined since all events are known to God simultaneously at all times. Yet Quantum tweaks make it possible for God to make events happen without violating Physical Laws (that God established). *All* events are recorded on the Tape of Fate *forever* from the "Beginning" (which is not truly a beginning since there is no beginning) to the end of eternity (which is not truly an end since there is no end).

Appendix G-F. God Theory's Related to Unified SuperStandard Theory

The God Theory presented in these editions was born in a study of the axioms of the Unified SuperStandard Theory of Elementary Particles[102] developed by the author. God Theory can be viewed as the source of the Unified SuperStandard Theory. On the other hand, the Unified SuperStandard Theory can be viewed as an independent theory that stands on its own merits.

We indicate the relation of God Theory to the Unified SuperStandard Theory and its other consequences in the following diagram.

[102] This appendix is inserted for the benefit of those who wish to have an idea of the relation of God Theory to the Unified SuperStandard Theory.

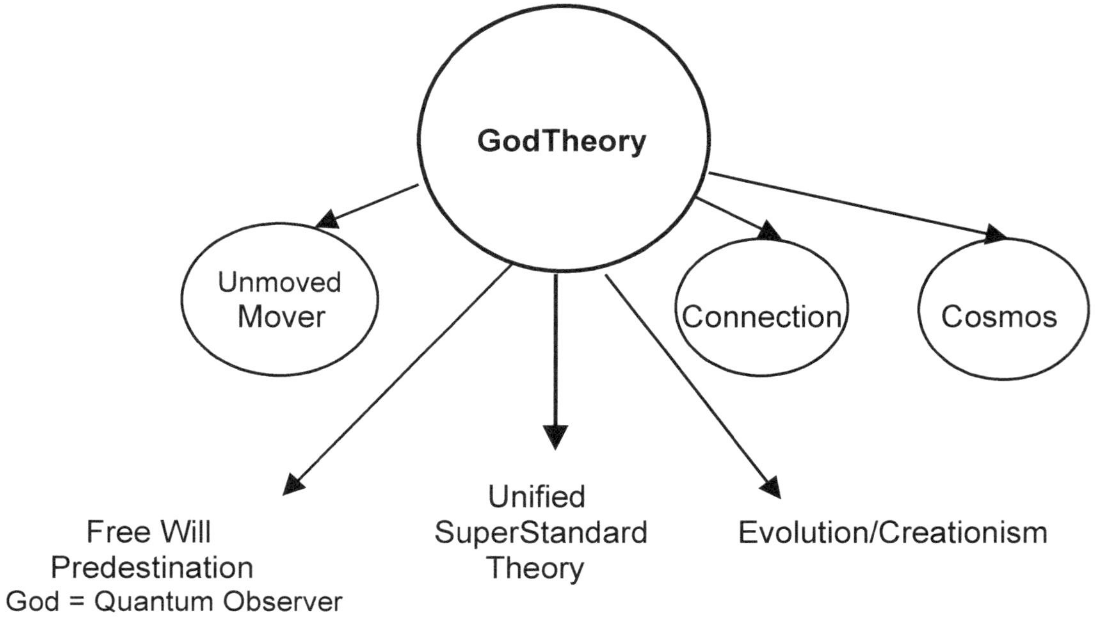

Appendix G-G. Possible Mechanism of the Connection and Cosmos Parts of God

G-G.1 Joining of Connection and Cosmos

The "union" of the Connection with the Cosmos is implemented by associating *two* quantum fields with every particle of matter and energy.[103] In Blaha (2018a) and (2018b) we implemented this formulation of the SuperStandard Theory for major technical reasons.

Based on our preceding discussions here, we now attribute a greater significance to the two field formulation. We identify one of the fields, in each pair of fields for a particle, as a Connection field that supports the roles of the Connection: 1) as forcing reality to conform to the Laws of Physics as set up by the Unmoved Mover, and 2) as implementing possible "Divine Intervention" by the Unmoved Mover. This field is typically labeled with a "1". The other field, which we label with the index "2", is a Cosmos field that generates the physical dynamics of Reality manipulating matter and energy. The dynamics equations are the result of the Connection fields in a lagrangian formulation.

In this book we have assumed that all particles of matter and energy have a spark of divinity within them that taken together makes both the Cosmos and Connection divine. Thus the fields of both type "1" and type "2" have a divine spark within them (when their particle relation is considered) by assumption—an assumption required in order to have a divine Cosmos and a divine Connection.

---

[103] Both boson and fermion particles.

G-G.2 Technical Reasons for "Two Fields per Particle" Formulation

The Unified SuperStandard Theory, and earlier work, pioneered the use of a quantum field theoretic formalism that associated two second quantized fields with every elementary particle. The two field formalism was also used to develop a generalization of Quantum Mechanics that united classical fields with quantum fields. (This formalism enabled a "continuous" transformation (rotation) to be made between strictly quantum fields and "corresponding" strictly classical fields.

In the Unified SuperStandard Theory we used the two second quantized fields' formalism in four situations:

1. The formalism was applied to Higgs boson fields to obtain a formulation that cleanly separated the vacuum expectation value of a Higgs particle's field from its second quantized part.

2. The formalism was applied to higher derivative dynamic parts to obtain a canonical lagrangian formulation via the Ostrogradski Bootstrap mechanism. The higher derivative sector generated quark confinement through a manifest linear potential, and a gravitational potential that had three forms depending on distance in a manner similar to MOND gravity. But without modifying fundamental Newtonian law.

3. The formalism resolves the well-known problem of ambiguities between second quantization in different reference frames. In the 1970's the author showed that the formalism could resolve second quantization ambiguities using Bogoliubov transformations. Second quantization then becomes "covariant" under changes of coordinate systems.

4. The two field formalism, if used in Quantum Mechanics, supports continuous transformations between quantum mechanics and classical mechanics, and allows for pseudo-quantum mechanics intermediate between quantum mechanics and classical mechanics. Such intermediate mechanics may appear in the "gray" zone that one sometimes sees in atomic theory.

G-G.3 Divine Cosmos and Divine Connection

The Connection is god-like through sparks of godliness within the fields of type "1" that are like the *qubes* and *qubas* of matter and energy in the universe (and the Megaverse) in our Unified SuperStandard Theory.

The Unified SuperStandard Theory specifies the dynamics of the divine Cosmos through the control on type "2" fields exerted by the type "1" fields of the Connection. Both Connection and Cosmos have Qubes and Qubas containing divine sparks.

The Unmoved Mover uses the Connection to constrain the Cosmos to obey Physical Laws and to possibly intervene in the Cosmos.

G-G.4 Unmoved Mover, Connection and Cosmos

Having linked the Cosmos and the Connection we now see the Unmoved Mover part of God to be the *ultimate source* (Origin) of Everything thus bringing our Unified Standard Theory to completion.

G-G.5 Fermion Spectrum of the Unified SuperStandard Theory

This appendix[104] describes a possible mechanism for the Connection based on our Unified SuperStandard Theory.[105,106] The fermion (spin ½) fundamental fermions in the Unified SuperStandard Theory are depicted in Fig. G-G.1. The possible newly found particles have a ? mark next to them in the Fermion Spectrum Table. We base our neutrino placement in the fourth generation on a heavier new neutrino mass, and the placement of a heavier (than an electron) electron-like particle in the second particles layer of the fermion (spin ½) particle spectrum due to its passage through the earth without interactions.

This section describes the "Periodic Table" of fermions (spin ½ fundamental particles like electrons, neutrinos, quarks and so on). It is analogous to the Periodic Table of Elements in Chemistry.

We choose to provisionally identify the e_2 particle as a second layer electron-like particle because it traversed the earth without creating an interaction shower. The interaction between particles in different layers can be expected to be ultraweak in the Unified SuperStandard Theory. (The subscript $_2$ indicates a second layer particle.)

The first layer neutrinos are then e, μ, τ, and the proposed υ – a fourth generation charged lepton.

The new particles are encouraging support for the Unified SuperStandard Theory.

[104] This appendix is inserted for the benefit of those who wish to have an idea of the relation of God Theory to the Unified SuperStandard Theory. The details appear in the second section of the book.

[105] Much of the material appeared in Blaha (2018a) and (2018b).

[106] The week of September 17, 2018 saw public announcements of a possible fourth generation neutrino seen at the Super-Kamiokande giant particle detector. If so, it would be a potential confirmation of the four fermion generations predicted by our Unified SuperStandard Model. Since then new particles have also been found that passed through the earth without generating an interaction shower. They may be second layer, charged leptons (heavier siblings of the electron.) See the paper D. B. Fox et al, arXiv:astro-ph.HE 1809.09615 (2018). The fact that these particles traverse earth without an interaction shower indicates they do not have known Standard Model interactions. Thus they are most likely members of a higher fermion layer. See Fig. G-G.1.

18. Monads of Elementary Particles – Extension to 10 Spaces

In previous books[107] the author has shown that the relativistic problem of instantaneous communication between parts of a quantum system at space-like distances can be resolved through the factoring of quantum wave functions into a functional and a spatial wave. Functionals, which embody the instantaneous effects, were in a point space of functionals. Spatial waves were in a point Fourier wave space. Thus instantaneous effects did not proceed through coordinate space distances. See Appendix 18-A.

In section G-6.2 of Chapter 17 we discussed the possibility that every particle has a monad within it that is in a sense part of God. In this chapter we show that monads explain instantaneous quantum entanglement by existing in a spaceless space with no distances within it. Since monads are part of individual's bodies, one can see that they exist after death leading to a sense of life after death. We refer the reader back to chapter 17 and provide a discussion of the impact of monads on life later. We turn now to the physics of monads.

The general form of a Dirac fermion wave function is

$$\psi(x) = \sum_{\pm s} \int d^3p N(p)[b(p, s)u(p, s)e^{-ip\cdot x} + d^\dagger(p, s)v(p, s)e^{+ip\cdot x}] \tag{18.1}$$

It can be "factored" into a functional inner product formalism in the manner of Riesz (1955)[108] and others:

$$\psi(x) = (f, (s, x, t)) \tag{18.2}$$

In the general case a functional for a single particle state can be expressed as a "product" of factors:

$$f_{IS,spin,momentum} = f_{IS}\, f_{spin,\ momentum}$$

where f_{IS} is a functional specifying the Internal Symmetry, and $f_{spin,\ momentum}$ is a functional specifying the spin and momentum.

We identify the functionals defined above as *monads*. In Blaha (2020c) and earlier books we showed that we can take every elementary particle to have a monad within it. Fermion monads, called *qubes*, appear to give a small mass to fermions. We

---

107 See for example Blaha (2020c) and references therein.

108 For example see pp. 61-2 of Riesz (1955) where linear functionals and their inner products are defined.

suggest this mass was of the order of the *Landauer energy*. Boson particle monad (*Quba)* masses were taken to be zero in order to avoid conflicts with gauge symmetries.

Justification for the "factoring" of wave functions into functionals and waves is presented in Appendix 18-B.

18.1 Internal Symmetry Functionals Space F_{IS}

We can view the internal symmetry functionals as resident in a space of functionals. This space has zero distance similar to the spaceless spaces considered in chapter 0. Such a space has a functional for each fermion and boson in QUeST and UST. This space can be extended to handle multi-particle states by supporting products and sums of individual particle functionals. It is infinite since it must support all possible particle configurations.

There is no distance in the set of functionals. Thus, as we showed in Blaha (2020c) and earlier books, factoring quantum fields into inner products of functionals and waves eliminates the instantaneity problem of Quantum Entanglement.

We will denote this point space as F_{IS}.

18.2 Spin-Momentum Functionals Space F_{SP}

We can define another infinite space consisting of functionals for all possible spin and momentum states. This space has no distance measure and can be viewed as a point space. We will denote this point space as F_{SP}.

18.3 Fourier Wave Space F_{FW}

We can define a space of all possible Fourier waves F_{FW}. This space also has no distance measure and can be viewed as a point space.

18.4 Space-Time and Functional Spaces

At each point in a universe's space-time, we can view the three point spaces above as present. They exist as a "tensor product" with the space-time. There is only one of each point space, which is shared by each space-time point.

A spatially dispersed quantum system has instantaneous access to its composite functional. *Thus a change in one part of the system causes instantaneous changes in other parts.*

18.5 Ten Spaces of Octonion Cosmology

The ten octonion spaces introduced in chapter 1, plus the three functional spaces presented here, gives us a thirteen Space Cosmology.

This chapter was presented within the framework of free fields and their products. It is general because all measurements are made on in-states and out-states. Quantum measurements within interaction regions are generally not possible.

18.6 Dynamic Generation of Quantum State Functionals

The preparation of a physical system of particles is always required to be local. After preparation, a quantum system can evolve to have widely separated parts at space-like distances.

The preparation of a quantum system requires functionals for each of the parts (particles) within it. These functionals are generated by dynamic extensions (and combinations) of the functionals of the three spaces defined above. Thus the above three spaces must be extendable—without limit—to accommodate all possible physical state preparations.

In the preparation of a state, the functionals of different parts of the state are interrelated leading to effects such as quantum entanglement.

18.7 Monads and The Possibility of a Form of Life After Death

In chapter 17 and above we consider monads as existing in all particles but also existing in a "point" space without distances. Since the matter in a person's body continues to exist after death, the monads within also continue to exist and continue to embody the content of a person's consciousness. Thus one can say that individuals continue to exist in a disembodied way after death. If a person's body is burned or otherwise dispersed over time the monads that were within, still persist but not necessarily in an organized way. The possibility that the monads with a body can cohere into an organized entity after death is not excluded.

A form of life after death is not inconceivable! Its basis can be understood within a Physics framework.

Appendix 18-A. Quantum Entanglement and Action-at-a-Distance

Einstein, Podolsky, and Rosen (EPR) considered the quantum entanglement of two systems and showed that instantaneous action-at-a-distance (spookiness) resulted. In this appendix we will show that our quantum functional formalism, which generalizes quantum theory, eliminates the problem of instantaneous action-at-a-distance.[109] We will show the solution provided by quantum functionals using the same example as EPR.

The key feature of quantum functionals is their ubiquitous presence at every point of space-time. In a multi-system quantum state all functionals are directly, instantaneously linked no matter what the separation of the constituent systems. When a reduction of part of the state of a system occurs due to a measurement, all other parts are instantly updated since the space-time separation of the individual parts is not relevant. The linkage of all quantum functionals is relevant. A reduction of one part 8.immediately impacts the other related parts.

18-A.1 The EPR Two System State Example

EPR considered a state consisting of two systems that might become separated spatially. We can represent the state as

$$\Psi = \Sigma_n \psi_{1n}(x_1)\psi_{2n}(x_2) \qquad (18\text{-A}.1)$$

We can represent a measurement (reduction of state) with a projection Π_{1a} of system "1" to a state ψ_{1a} with

$$\psi_{1a} = \delta_{ab}\, \Pi_a\, \psi_{1b} \qquad (18\text{-A}.2)$$

Then

$$\Psi_{\text{projected}} = \Pi_{1a}\, \Sigma_n \psi_{1n}\psi_{2n} = \psi_{1a}(x_1)\psi_{2a}(x_2) \qquad (18\text{-A}.3)$$

The effect of the measurement of system "1" is *instantaneous* on system "2" because the quantum functionals f_{1n} and f_{2n}, and the projections Π_{1n} and Π_{2n} of both systems are not separated by distance. Thus

$$\psi_{1n}(x) = f_{1xn}(\Pi_{1xn}\Phi\,) = (f_{1xn}, \Pi_{1xn}\Phi) \qquad (18\text{-A}.4)$$
$$\psi_{2n}(y) = f_{2ny}(\Pi_{2yn}\Phi\,) = (f_{2yn}\,, \Pi_{2yn}\Phi) \qquad (18\text{-A}.5)$$

[109] This section presents the solution for quantum spookiness that we proposed in Blaha (2019g) and (2018e).

with $x = x_1$ and $y = x_2$. *The quantum functional and the projection select the wave and its coordinate parameterization. The coordinates in the wave are merely place holders.*

Therefore the relative distance between the coordinates x_1 and x_2 is not relevant for the change of state of system "2". The quantum functionals and projections give the instantaneity of the change in ψ_{2a} upon the measurement of system "1".

EPR Spookiness is resolved by quantum functionals. There is no conflict with the Theory of Special Relativity.

Appendix 18-B. Why Factorize Quantum Fields into a Functional and a Fourier Expansion?

We identify three reasons for factoring quantum fields:

1. The internal symmetries of free fundamental fermions are only partially correlated with their fourier expansions: there is a spin/handedness relation and a species relationship based on the species of each fundamental fermion. The internal symmetries: color, generation number, layer number, and so on appear only as indices attached to creation and annihilation operators. Functionals are matched with corresponding fourier wave expansions using functional inner products.

2. Quantum Entanglement between parts of a physical state separated by space-like distances suggest the correlated changes in the quantum numbers of a physical state with spatially separated parts are independent of spatial distance. Quantum number changes (in any quantum numbers) between spatially separated parts are instantaneous. This fact demonstrates phenomena strongly supports the factorization of the quantum numbers and the coordinates of quantum fields.

3. The close analogy between the subgroups of the Complex Lorentz group and the known symmetries of The Standard Model suggest that the internal symmetries of fermions represented by functionals have a somewhat similar group structure to space-time symmetries.

For these reasons we propose to consider quantum fields as inner products of functionals and space-time fourier waves.

19. Octonion Cosmology

This volume presents the panorama of Octonion and Functional Spaces needed for a Physical Cosmology at the deepest level of Physical Reality. Fig. 19.1 diagrams the 10 spaces and their instances. It is difficult to envision a deeper, comprehensive, physical level of Cosmology due to its basis in Number.

As seen earlier, Octonion Cosmology leads to the Unified SuperStandard Theory (UST) which the author developed in the past twenty years, that is based on the methods of Euclid's Logic. UST contains the Standard Model of Elementary Particles within the framework of an extended Quantum Field Theory that originates in the author's papers of the 1970's.

The beauty of the octonion spectrum of spaces presented in Fig. 1.4 argues for its elegance and simplicity as well as its fundamental significance. The directness of the derivation of UST, and its close connection to experimental reality, as far as we know it, suggests it has the general framework needed to understand elementary particle physics in detail.

The correspondence of Octonion Cosmology with the Cosmology of the *Sefer Yetzirah* encourages the hope that Physics is reaching the deepest Philosophic/Religious level of Reality. This correspondence leads to the hope of an eventual convergence of Physics, Philosophy and Religion.

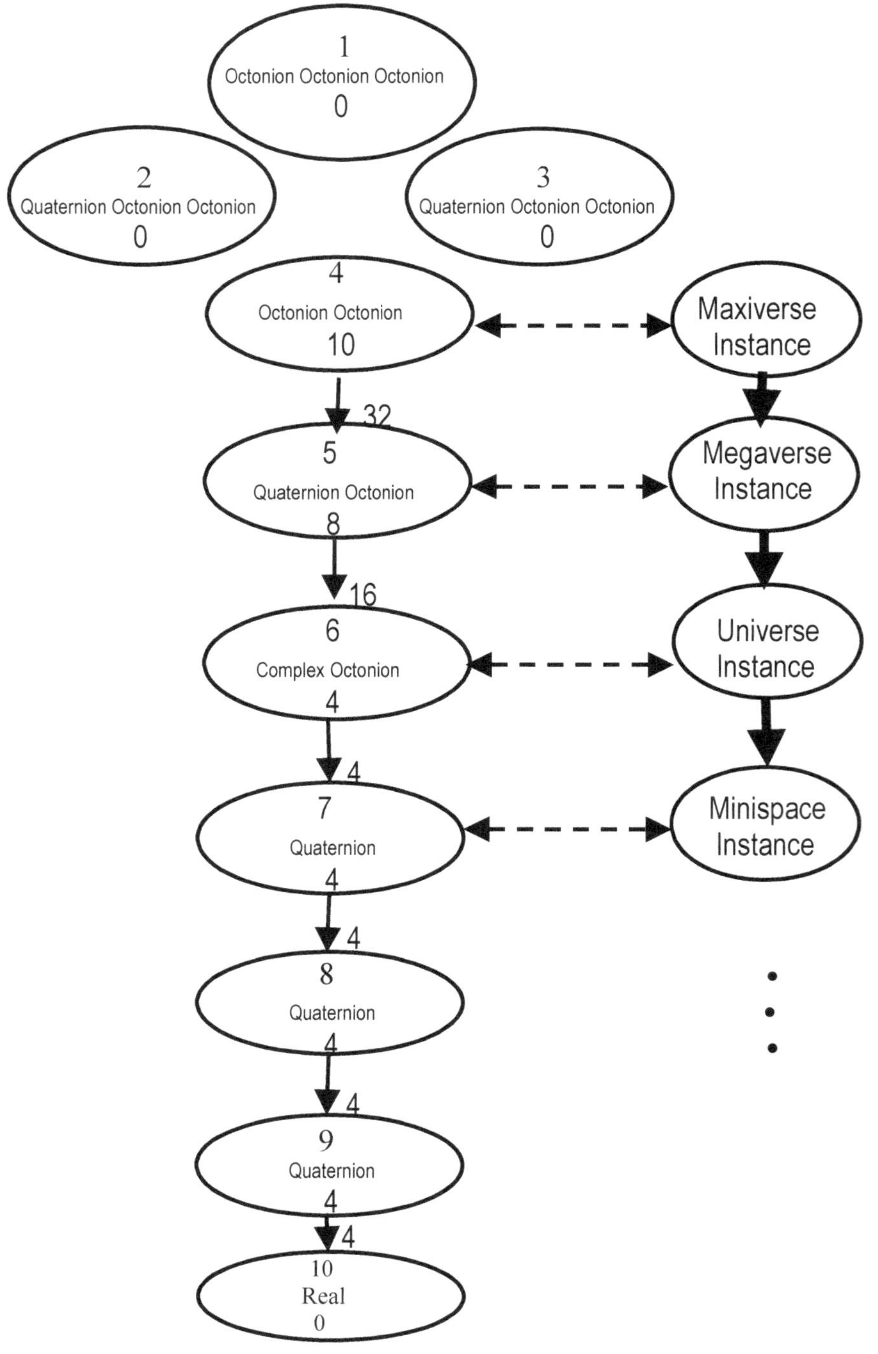

Figure 19.1. The ten spaces with their spectrum number and their space-time dimensions indicated within each oval as in Fig. 1.4. The number of spinor components for each fermion - antifermion pair that annihilates to produce the "next space down" is specified next to each arrow for the lower spaces. On the right are symbols for instances (particles) of spaces.

Appendix A. Enhanced Quantum Field Theory: Two Tier & PseudoQuantum from Chapters 40 and 41 of Blaha (2020c)

This Appendix describes Two-Tier Quantum Coordinates (Chapter 40 below) and PseudoQuantum field Theory (Chapter 41 below).

We begin by outlining the advantages of PseudoQuantum Field Theory (from Blaha (2020c) and earlier books). To these advantages we add the capability of supporting the derivation of QUeST from BQUeST and the derivation of UTMOST from BMOST. The advantages are:

1. Quantization in any coordinate system in flat or curved space-times with an invariant definition of asymptotic particle states. An n particle asymptotic state in one coordinate system is a unitarily equivalent n particle asymptotic state in any other coordinate system. Therefore particle number is invariant under change of coordinate system. This is important for the Unified SuperStandard Theory in curved space-times. It is also important for quantization in higher dimensional Euclidean spaces such as the Megaverse. The method was developed in the late 1970's by the author to provide a quantization procedure which supports a unique particle interpretation of states in arbitrary non-static space-times where no global timelike coordinate (Killing vector) exists. PseudoQuantum Field Theory which we developed in a series of books[110] also can be formulated in the Megaverse. Thus we can use it in the Megaverse to implement the Higgs Mechanism to generate particle masses and symmetry breaking.

2. PseudoQuantum Field Theory enables one to define Higgs particle dynamics in such a way that a non-zero vacuum expectation value cleanly separates from the quantum field part of the Higgs fields. This technique can be used in symmetry breaking mechanisms, mass generation, and possible generation of coupling constants as vacuum expectation values.

3. It supports the canonical definition of higher derivative field theories through the use of the Ostrogradski bootstrap. See Appendix B where a fourth order theory of the Strong interaction is defined that has color confinement and a linear r potential. The potential part of this theory was used by the Cornell group to calculate the Charmonium spectrum. (See Blaha (2017b) for details.)

---

[110] See Blaha (2017b) for the discussion of the PseudoQuantum field theory formalism for Higgs particles in our Extended Standard Model. See chapter 20 of Blaha (2017b), and earlier books, for a more detailed view than that presented here.

An associated advantage of using PseudoQuantum Field Theory is that it provides for retarded propagators and an Arrow of Time.
AND

4. Support for derivations of QUeST and UTMOST from one dimension BQUeST and BMOST.

See the author's paper, S. Blaha, Il Nuovo Cimento **49A**, 35 (1979). Other papers from the 1970's are listed therein. Blaha (2007b) and (2011c) describe PseudoQuantum Field theory.

Blaha (2016f) describes PseudoQuantum Mechanics (*CQ Mechanics*), a related formalism, and describes applications (including the Schrödinger equation, Fokker-Planck equation, Boltzmann equation, and the Vlasov equation) that smoothly provide *continuous* transitions between Quantum Theory and Classical Theory.

We now turn to describing Two-Tier Quantum Coordinates.

40. Two-Tier Coordinates

Originally Two-Tier coordinates were developed by this author to remove infinities that appear in perturbation theory calculations. We showed that the quantum smeared coordinates of Two-Tier Quantum Field Theory succeeded in removing all ultra-violet infinities in perturbation theory including the fermion triangle infinities. Remarkably the high precision, low energy[111] predictions of QED remained true in Two-Tier QED and thus remained consistent with experiment to a hitherto unsurpassed level of accuracy. 'Low' energy predictions in other quantum field theories also remained unchanged. At high energies, Two-Tier perturbation theory results are finite and consequently all ultra-violet infinities, to any order in perturbation theory, in *any number of space-time dimensions* were eliminated.

In addition to removing perturbation theory infinities Two-Tier coordinates enable us to define finite theories of Quantum Gravity and 'non-renormalizable' quantum field theories based on polynomial lagrangians, to tame vacuum fluctuations, to eliminate infinities associated with the Big Bang, and possibly to generate the explosive growth of the universe in its role as Dark Energy.[112]

Two-Tier Quantum Field Theory is established on the most fundamental level.

40.1 Two-Tier Features in 4-Dimensional Space-Time

Two-Tier Quantum Field Theory,[113] which was based on a new method[114] in the Calculus of Variations, uses two sets of fields to introduce quantum coordinates. We

---

[111] Relative to a mass scale that was perhaps of the order of the Planck mass.
[112] See Blaha (2017b) and earlier books for details. This section is basically a summary of some features.
[113] See Blaha (2005a), and Blaha (2002), for discussions of this new method to eliminate infinities in quantum field theory calculations.
[114] Blaha (2005a) describes our method for the composition of extrema in some detail.

shall consider this technique for the specific case of a massless vector field $V^i(y)$ analogous to the electromagnetic field.

In 4-dimensional space-time the massless vector field has the form $Y^\mu(y)$ where the index μ ranges from 0 through 3. The x^μ coordinate system, where it appears, has a c-number real part and a q-number imaginary part. Thus particle fields which are normally defined on four-dimensional real space-time will now be defined on a complex four-dimensional space-time where four imaginary dimensions will appear as *Quantum Dimensions* embodied in a vector quantum field $Y^\mu(y)$:

$$X^\mu(y) = y^\mu + i\, Y^\mu(y)/M_c^{\,2}$$

where M_c is an extremely large mass of the order of the Planck mass or perhaps much larger.

The $Y^\mu(y)$ field is a function of the subspace y coordinates. The real part of the space-time dimensions will be taken to be the space of real-valued y coordinates.[115]

The imaginary part of space-time coordinates is the a massless $Y^\mu(y)$ vector quantum field that is suppressed further by a very large mass scale – perhaps of the order of the Planck mass – that reduces the imaginary Quantum Dimensions to the infinitesimal except at large momenta. The effects of Quantum Dimensions only become appreciable in quantum field theory at energies of the order of M_c. At these energies exponential Gaussian factors in each particle (and ghost) propagator are generated by the Quantum Dimensions and serve to make perturbation theory calculations ultra-violet finite – including calculations in Quantum Gravity.

The formalism introduces a new form of interaction that does not have the form of the simple polynomial interactions that have hitherto dominated quantum field theories. This form of interaction takes place via the composition of quantum fields and can be called a *Dimensional Interaction,* or an *Interdimensional Interaction,* since it affects particle behavior through Quantum Dimensions.

The basic ansatz of the Two-Tier formalism is to replace every appearance of a coordinate x in a quantum field with the variable

$$x^\mu \rightarrow X^\mu = (y^0, \mathbf{y} + \mathbf{Y}(y^0, \mathbf{y})/M_c^{\,2})$$

where $\mathbf{Y}(y^0, \mathbf{y})$ is the spatial part of a free massless vector field with features that are identical to the free QED field in the Radiation gauge.

Then one finds that the momentum space free field Feynman propagators G(k) of all particles acquires a Gaussian factor exp(h(k)):

$$G(k) \rightarrow G(k)\, \exp(h(k))$$

---

[115] In a deeper theory the real part might also be a quantum field that undergoes a condensation to generate c-number coordinates. We will not consider this possibility in this book.

so that all perturbation theory diagrams are finite. The result is finite perturbative results for all calculations to any order in perturbation theory. Blaha (2005a) shows that Two-Tier theories are finite, Poincare covariant, and unitary. (See Blaha (2005a), chapter 5, for a complete discussion.)

40.2 Simple Two-Tier X^μ Formalism

In this subsection we will describe the basic Two-Tier formalism. Taking the lagrangian described in Blaha (2005a):[116]

$$\mathscr{L}(y) = \mathscr{L}_F(X^\mu(y))J + \mathscr{L}_C(X^\mu(y), \partial X^\mu(y)/\partial y^\nu, y) \tag{40.1}$$

where

$$X^\mu(y) = y^\mu + i\, Y^\mu(y)/M_c^2 \tag{40.2}$$

with M_c being a large mass scale, $Y_\mu(y)$ a vector quantum field, and where J is the absolute value of the Jacobian of the transformation from X to y coordinates:

$$J = |\partial(X)/\partial(y)|$$

The lagrangian term $\mathscr{L}_C$ is

$$\mathscr{L}_C = +\tfrac{1}{4}\, M_c^4 F^{\mu\nu}F_{\mu\nu}$$

with

$$\begin{aligned} F_{\mu\nu} &= \partial X_\mu/\partial y^\nu - \partial X_\nu/\partial y^\mu \\ &\equiv i\,(\partial Y_\mu/\partial y^\nu - \partial Y_\nu/\partial y^\mu)/M_c^2 \end{aligned} \tag{40.3}$$

The lagrangian term $\mathscr{L}_F$ $(X^\mu(y))$ contains the terms for scalar, fermion and other gauge terms in general. The sign in $\mathscr{L}_C$ is not negative – contrary to the conventional electromagnetic Lagrangian. The reason for this difference is that the quantum field part of X^μ is imaginary. Thus $\mathscr{L}_C$ ends up having the correct sign after taking account of the factor of i in the field strength $F_{\mu\nu}$.

Defining

$$F_{Y_{\mu\nu}} = (\partial Y_\mu/\partial y^\nu - \partial Y_\nu/\partial y^\mu)$$

we see the Lagrangian assumes the form of the conventional electromagnetic Lagrangian:

$$\mathscr{L}_C = -\tfrac{1}{4}\, F_Y^{\mu\nu}F_{Y_{\mu\nu}}$$

The action of this theory has the form

$$I = \int d^4y\, \mathscr{L}(y)$$

[116] Eq. 7.1.

40.3 Y^μ Gauge

The gauge invariance of the Lagrangian allows us to choose a convenient gauge. The gauge invariance of the full Lagrangian

$$\mathscr{L}_s = L_F(\phi(X), \partial\phi/\partial X^\mu)\, J + \mathscr{L}_C(X^\mu(y), \partial X^\mu(y)/\partial y^\nu)$$

is based on the standard gauge invariance of $\mathscr{L}_C$, and the gauge invariance of $J\mathscr{L}_F$ in the form of translational invariance

$$X^\mu(y) \rightarrow X^\mu(y) + \delta X^\mu(y)$$

for the special case of a translation of X with the form of a gauge transformation:

$$\delta X^\mu(y) = \partial\Lambda(y)/\partial y_\mu$$

In this case we find

$$\int d^4y\ \Lambda(y)\, \partial\, [\, J\, \partial/\partial X^\mu\, \mathscr{T}_{F\mu\nu}\,]/\partial y_\nu\ = 0 \tag{40.4}$$

after a partial integration and so we have the differential conservation law:

$$\partial\, [\, J\, \partial \mathscr{T}_{F\mu\nu}/\partial X^\mu]/\partial y_\nu\ = 0$$

since $\Lambda(y)$ is arbitrary. This conservation law is trivially obeyed:

$$\partial \mathscr{T}_{F\mu\nu}/\partial X^\mu = 0 \tag{40.5}$$

Thus translational invariance in the $\mathscr{L}_F$ sector together with standard gauge invariance in the $\mathscr{L}_C$ sector automatically guarantees Y field gauge invariance of the total Lagrangian. We use the separate invariance of each term of

$$L = \int d^4y\ [\mathscr{L}_F\, J + \mathscr{L}_C\,] = \int d^4X\ \mathscr{L}_F\ + \int d^4y\ \mathscr{L}_C = L_F + L_C$$

under a constant translation $X^\mu \rightarrow X^\mu + \delta X^\mu$ where δX^μ is constant. Then we consider a position dependent translation/gauge transformation, which taken together with the above equation, establishes the invariance under the position dependent translation/gauge transformation.

An alternate approach that leads to the same result is to start with the particle part of the Lagrangian $\mathscr{L}_F$ rewritten to be invariant under general coordinate transformations, as it must, when we generalize to include General Relativity. Since position dependent translations are a form of general coordinate transformation the full theory must be invariant under position dependent translations due to invariance under general coordinate transformations.

Having established invariance under gauge transformations we now choose to use the most convenient gauge – the radiation gauge[117]:

$$\partial Y^i/\partial y^i = 0 \tag{40.6}$$

where i = 1, 2, 3, which, in the absence of external sources, allows us to set

$$Y^0 = 0$$

since Y^0 does not have a canonically conjugate momentum. A conventional treatment leads to the equal time commutation relations:

$$[Y^\mu(\mathbf{y}, y^0), Y^\nu(\mathbf{y'}, y^0)] = [\pi^\mu(\mathbf{y}, y^0), \pi^\nu(\mathbf{y'}, y^0)] = 0 \tag{40.7}$$

$$[\pi^j(\mathbf{y}, y^0), Y_k(\mathbf{y'}, y^0)] = -i\,\delta^{tr}{}_{jk}(\mathbf{y} - \mathbf{y'})$$

(Note the locations of the j indexes above introduce a minus sign.) where

$$\pi^k = \partial\mathscr{L}_C/\partial Y_k'$$
$$\pi^0 = 0$$

$$\delta^{tr}{}_{jk}(\mathbf{y} - \mathbf{y'}) = \int d^3k\, e^{i\,\mathbf{k}\cdot(\mathbf{y} - \mathbf{y'})}(\delta_{jk} - k_j k_k/\mathbf{k}^2)/(2\pi)^3$$

$$Y_k' = \partial Y_k/\partial y^0$$

The Radiation gauge reveals the two degrees of freedom that are present in the vector potential. The Fourier expansion of the vector potential is:

$$Y^i(y) = \int d^3k\, N_0(k) \sum_{\lambda=1}^{2} \varepsilon^i(k, \lambda)[a(k,\lambda)\, e^{-ik\cdot y} + a^\dagger(k,\lambda)\, e^{ik\cdot y}] \tag{40.8}$$

where

$$N_0(k) = [(2\pi)^3 2\omega_k]^{-1/2}$$

and (since m = 0)

$$\omega_k = (\mathbf{k}^2)^{1/2} = k^0$$

---

[117] It is also possible to quantize using an indefinite metric that preserves manifest Lorentz covariance as was done by Gupta and Bleuler for the electromagnetic field. We will use the Gupta-Bleuler approach later to establish covariance under special relativity later. Now we opt for manifest positivity and use the radiation gauge.

with $\vec{\epsilon}(k, \lambda)$ being the polarization unit vectors for $\lambda = 1,2$ and $k^\mu k_\mu = 0$.

The further development of this theory is described in Part 3 of Blaha (2005a).

40.4 Scalar Field Quantization Using X^μ

We will begin by considering the case of a scalar quantum field theory. We assume a real underlying y subspace. Since X^μ is a set of coordinates, we choose to define a scalar field ϕ as a function of X^μ, which, in turn, is a function of the y^ν coordinates. We will provisionally second quantize ϕ treating X^μ as c-number coordinates using a conventional approach.[118]

We assume a Lagrangian, with the momentum conjugate to ϕ:

$$\pi_\phi = \partial L_F / \partial\phi' \equiv \partial L_F / \partial(\partial\phi/\partial X^0) \qquad (40.9)$$

Following the canonical quantization procedure, π and ϕ become hermitian operators with equal time ($X^0 = X^{0\prime}$) commutation rules:

$$[\phi(X), \phi(X')] = [\pi_\phi(X), \pi_\phi(X')] = 0 \qquad (40.10)$$
$$[\pi_\phi(X), \phi(X')] = -i\,\delta^3(\mathbf{X} - \mathbf{X'})$$

The standard Fourier expansion of the solution to the Klein-Gordon equation is:

$$\phi(X) = \int d^3p\, N_m(p)\, [a(p)\, e^{-ip\cdot X} + a^\dagger(p)\, e^{ip\cdot X}]$$

where

$$N_m(p) = [(2\pi)^3 2\omega_p]^{-1/2}$$

and

$$\omega_p = (\mathbf{p}^2 + m^2)^{1/2}$$

The commutation relations of the Fourier coefficient operators are:

$$[a(p), a^\dagger(p')] = \delta^3(\mathbf{p} - \mathbf{p'})$$
$$[a^\dagger(p), a^\dagger(p')] = [a(p), a(p')] = 0$$

The reader will recognize the quantization procedure is formally identical to the standard canonical quantization procedure of a free scalar quantum field.

---

[118] Some texts are: Bogoliubov, N. N., Shirkov, D. V., *Introduction to the Theory of Quantized Fields* (Wiley-Interscience Publishers Inc., New York, 1959); Bjorken, J. D., Drell, S. D., *Relativistic Quantum Fields* (McGraw-Hill, New York, 1965); Huang, K., *Quarks, Leptons & Gauge Fields Second Edition* (World Scientific, River Edge, NJ, 1992); Kaku, M., *Quantum Field Theory* (Oxford University Press, New York, 1993); Weinberg, S., *The Quantum Theory of Fields* (Cambridge University Press, New York, 1995).

In the case of spin ½, spin 1 and spin 2 fields the standard quantization procedure *in terms of the X coordinate system* can also be followed in a way similar to the procedure in standard texts.

40.5 Scalar Feynman Propagators

The momentum space free field Feynman propagators G…(k) of all particles and ghosts in all Two-Tier Quantum Field Theories acquires a Gaussian factor exp(h(k)):

$$G\ldots(k) \rightarrow G\ldots(k)\exp(h(k))$$

so that all perturbation theory diagrams are finite. The result is a finite perturbative result in all calculations to any order in perturbation theory. Blaha (2005a) shows that Two-Tier theories are finite, Poincare covariant, and unitary.

An example of the Two-Tier effect on propagators is the case of the Two-Tier photon propagator. The Two-Tier photon propagator[119] is:

$$iD_F^{TT}(y_1 - y_2)_{\mu\nu} = -i \int \frac{d^4p\, e^{-ip\cdot z}\, g_{\mu\nu} R(\mathbf{p}, z)}{(2\pi)^4\, (p^2 + i\varepsilon)} \tag{40.11}$$

(since the imaginary parts can be taken to be zero: $y_{1i}^{\mu} - y_{2i}^{\mu} = 0$) where

$$z^{\mu} = y_{1r}^{\mu} - y_{2r}^{\mu}$$

$$R(\mathbf{p}, \mathbf{z}) = \exp[-p^i p^j \Delta_{Tij}(z)/M_c^4]$$

$$= \exp\{ -\mathbf{p}^2[A(v) + B(v)\cos^2\theta] \,/\, [4\pi^2 M_c^4|\mathbf{z}|^2$$

with i, j = 1, 2, 3, and with $\Delta_{Tij}(z)$ being the commutator of the positive frequency part $Y^+{}_k(y)$ and the negative frequency part $Y^-{}_k(y)$ of $Y_k(y)$:

$$\Delta_{Tij}(z) = [Y^+{}_j(y_{1r}), Y^-{}_k(y_{2r})] = \int d^3k\, e^{ik\cdot(y_{1r} - y_{2r})}\, (\delta_{jk} - k_j k_k/\mathbf{k}^2)/[(2\pi)^3 2\omega_k] \tag{40.12}$$

and

$$v = |z^0|/|\mathbf{z}|$$
$$A(v) = (1 - v^2)^{-1} + .5v\ln[(v-1)/(v+1)]$$
$$B(v) = v^2(1 - v^2)^{-1} - 1.5v\ln[(v-1)/(v+1)]$$
$$\mathbf{p}\cdot\mathbf{z} = |\mathbf{p}|\, |\mathbf{z}| \cos\theta$$

---

[119] Blaha (2005a).

with $|\mathbf{p}|$ denoting the length of a spatial vector **p**, $|\mathbf{z}|$ denoting the length of a spatial vector **z**, and with $|z^0|$ being the absolute value of z^0.

The Gaussian factors R(**p,** z) which appear in all Two-Tier propagators damp the large momentum behavior of all perturbation theory integrals producing a completely finite perturbation theory and yet give the usual results of perturbation theory at energies that are small compared to the mass scale M_c.

40.6 String-like Substructure of the Theory

Two-tier Quantum field Theory endows each particle with an extended structure that resembles the extended structure seen in boson string and Superstring theories. For example, Bailin (1994) use the operator[120]

$$V_\Lambda(k) = \int d^2\sigma \sqrt{-h}\, W_\Lambda(\tau, \sigma)\, e^{-ik\cdot X}$$

where X^μ is a quantized fourier expansion of the string fields (see eq. 7.22 of Bailin (1994)).

We note our X^μ coordinate-field has two transverse degrees of freedom due to gauge invariance, which also invites comparison to the boson string. A point of difference is that we have a well-defined quantum field theoretic formulation in conventional space-time that has the Standard Model as its "large distance" behavior thus introducing a note of reality that is not apparent in Superstring theories. We see that the interacting quantum field theories based on this approach also have good, finite, short distance behavior just as string theories.

The scalar, and other particles', Feynman propagators can be viewed as describing the propagation of a particle cloaked (accompanied) by a cloud of Y particles (which generates the R(**p,** $y_1 - y_2$) factor in the above propagator). If we examine the fourier transform of R(p, z) we see:

$$(2\pi)^4 R(\mathbf{p}, q) = \int d^4z\, e^{iq\cdot z}\, R(\mathbf{p}, z) = \int d^4z\, e^{iq\cdot z} \exp[-p^i p^j \Delta_{Tij}(z)/M_c^4] \qquad (40.13)$$

and we find

$$R(\mathbf{p},q) = \sum_{n=0}^{\infty} [i(2\pi M_c)^4]^{-n} (n!)^{-1} \prod_{j=1}^{n} [\int d^4k_j\, \theta(k_j^0)(\mathbf{p}^2 - (\mathbf{p}\cdot\mathbf{k}_j)^2/\mathbf{k}_j^2)/(k_j^2 + i\varepsilon)]\, \delta^4(q - \sum k_r)$$

which can be interpreted as a "cloud" of Y particles dressing the "bare" particle propagator. (The apparent divergences for R(p, q) are an artifact of the expansion and the subsequent fourier transformation. They are not present in the R(**p,** $y_1 - y_2$) factor in the propagator. See Fig. 40.1 for the Feynman diagram of the Two-Tier 'cloaked' propagator as compared to the normal scalar particle Feynman propagator. The Two-

---

[120] D. Bailin and A. Love, *Supersymmetric Gauge Field Theory and String Theory* (Institute of Physics Publishing, Philadelphia, PA, 1994) page 272.

Tier Feynman propagator is basically a conventional scalar propagator that is modified by coherent Y particle emission.[121]

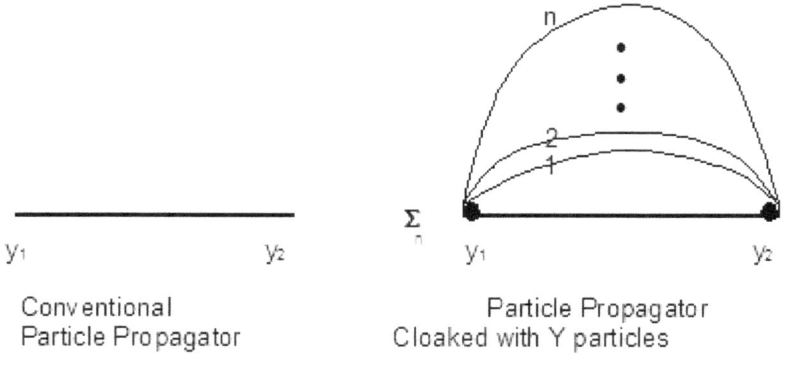

Figure 40.1. Feynman diagram for conventional and the nth diagram of a cloaked Two-Tier propagator.

We note that R(p, q) satisfies the convolution theorem:

$$\int d^4k\, R(\mathbf{p}, k)\, R(\mathbf{p}, q - k) = [R(\mathbf{p}, q)]^2$$

or

$$(2\pi)^4 \int d^4z\, e^{iq\cdot z}\, R(\mathbf{p}, z)\, R(\mathbf{p}, z) = [\int d^4z\, e^{iq\cdot z}\, R(\mathbf{p}, z)\,]^2 \qquad (40.14)$$

The proof follows from the Binomial theorem.

40.7 Two-Tier Complexon Quantum Fields

In the case of the Complexon Standard Model we will need two variables X_r^μ and X_i^μ since we have complex spatial 3-coordinates. We define them similarly to the previous case:

$$X_r^\mu(y_r) = y_r^\mu + i\, Y_r^\mu(y_r)/M_c^2$$
$$X_i^\mu(y_i) = y_i^\mu + i\, Y_i^\mu(y_i)/M_c^2$$

where we choose the same mass scale for both the "real" and "imaginary" variables. The Two-Tier, single generation, version of the Complexon Standard Model then has an action of the form

$$I_{CSMtt} = \int dy^0 d^3y_r d^3y_i\, (\mathscr{L}_{CSM}(X_r^\mu(y_r), \mathbf{X}_i^k(y_i))J_2)|_{yi0 = 0,\, Yr0 = Yi0 = 0} + \qquad (40.15)$$
$$+ \int dy_r^0 d^3y_r\, \mathscr{L}_C(X_r^\mu(y_r), \partial X_r^\mu(y_r)/\partial y_r^\nu, y_r) +$$

[121] T. W. B. Kibble, Phys. Rev. **173**, 1527 (1968) and references therein. In particular see p. 1532 of Kibble's paper.

$$+ \int dy_i^0 d^3y_i \mathscr{L}_C(X_i^\mu(y_i), \partial X_i^\mu(y_i)/\partial y_i^\nu, y_i)$$

where the replacements

$$x^\mu \equiv x_r^\mu \rightarrow X_r^\mu(y_r)$$
$$x_i^k \rightarrow X_i^k(y_i)$$

for $\mu = 0, 1, 2, 3$ and $k = 1, 2, 3$ are made, followed by defining $y_r^0 = y^0$ and making a Complex Lorentz transformation to a frame where $y_i^0 = 0$. J_2 is the absolute value of the Jacobian of the transformation from (X_r, X_i) to (y_r, y_i) coordinates:

$$J_2 = |\partial(X_r, X_i)/\partial(y_r, y_i)|$$

We also choose gauges where $Y_r^0 = Y_i^0 = 0$. These types of transformations and gauge choices are discussed in detail in Blaha (2005a). The lagrangian terms $\mathscr{L}_C(X_r^\mu(y_r), \partial X_r^\mu(y_r)/\partial y_r^\nu, y_r)$ and $\mathscr{L}_C(X_i^\mu(y_i), \partial X_i^\mu(y_i)/\partial y_i^\nu, y_i)$ have the same form:

$$\mathscr{L}_C = +¼\, M_c^4 F^{\mu\nu}F_{\mu\nu} \tag{40.16}$$

with

$$F_{\mu\nu} = \partial X_\mu/\partial y^\nu - \partial X_\nu/\partial y^\mu$$
$$\equiv i\,(\partial Y_\mu/\partial y^\nu - \partial Y_\nu/\partial y^\mu)/M_c^2$$

or defining

$$F_{Y\mu\nu} = (\partial Y_\mu/\partial y^\nu - \partial Y_\nu/\partial y^\mu)$$

we see each lagrangian assumes the form of the conventional electromagnetic Lagrangian:

$$\mathscr{L}_C = -¼\, F_Y^{\mu\nu}F_{Y\mu\nu}$$

The lagrangian is supplemented with the following condition on all complexon fields $\Phi_{...}$:

$$(\partial/\partial X_r^k(y_r))\,(\partial/\partial X_i^k(y_i))\Phi\ldots = 0 \tag{40.17}$$

summed over $k = 1, 2, 3$. Non-complexon fields $\Omega\ldots$ in our left-handed formulation satisfy the subsidiary condition:

$$\{(\partial/\partial X_r^k(y_r))(\partial/\partial X_i^k(y_i)) - [(\partial/\partial X_r^k(y_r))^2(\partial/\partial X_i^m(y_i))^2]^{½}\}\Omega\ldots = 0 \tag{40.18}$$

summed over $k = 1, 2, 3$ and over $m = 1, 2, 3$ separately in each of the two terms.

40.8 Complexon Feynman Propagator

In the case of complexons, the Two-Tier Feynman propagator differs from the non-complexon case by having an integration over imaginary spatial 3-momenta, a derivative of a delta function embodying the orthogonality of the real and imaginary 3-momenta, and two factors of R(**p**, z): one factor being R($\mathbf{p_r}$, $\mathbf{z_r}$) and the other factor being R($\mathbf{p_i}$, z_i) (where the time components $z_r^0 = z^0$ and $z_i^0 = 0$ since there is only one real time coordinate[122]) thus providing large momentum convergence for both real and imaginary 3-momentum integrations.

For a normal scalar particle the Feynman propagator is:

$$i\Delta_{CTF}(x-y) = \theta(x^+ - y^+)<0|\phi_{CT}(x)\,\phi_{CT}(y)|0> + \theta(y^+ - x^+)<0|\phi_{CT}(y)\phi_{CT}(x)|0>$$
$$= i\int d^4p_r d^3p_i (2\pi)^{-7}\delta'(\mathbf{p_r}\cdot\mathbf{p_i}/m^2)e^{-ip^+(x^- - y^-) - ip^-(x^+ - y^+) + ip_\perp\cdot(x_\perp - y_\perp) - ip_i\cdot(x_i - y_i)}/(p^2 + m^2 + i\varepsilon) \quad (40.19)$$

in conventional quantum field theory.

In the case of Two-Tier quantum field a scalar *complexon* particle has the Feynman propagator

$$i\Delta_{CTFtt}(x-y) = i\int d^4p_r d^3p_i (2\pi)^{-7}\delta'(\mathbf{p_r}\cdot\mathbf{p_i}/m^2)\, R(\mathbf{p}_r, z_r)R(\mathbf{p}_i, z_i)\cdot \quad (40.20)$$
$$\cdot e^{-ip^+(x^- - y^-) - ip^-(x^+ - y^+) + ip_\perp\cdot(x_\perp - y_\perp) - ip_i\cdot(x_i - y_i)}/(p^2 - m^2 + i\varepsilon)$$

where the time components $z_r^0 = z^0$ and $z_i^0 = 0$ since there is only one time coordinate, where R(**p**, z) is given in the previous subsection, and where $p^2. = p^{0\,2} - p_r^2 + p_i^2$.

Propagators for other types of particles are similarly modified in the Two-Tier formalism (See Blaha 2005a).

40.9 Vacuum Fluctuations

While the expectation value of a *conventional* free scalar field $\phi_{conv}(x)$ is zero in a conventional quantum field theory:

$$<0|\phi_{conv}(x)|0> = 0 \quad (40.21)$$

the vacuum fluctuations of *conventional* scalar quantum field theory are quadratically divergent:

$$<0|\phi_{conv}(x)\phi_{conv}(x)|0> = \int d^3p/[(2\pi)^3 2\omega_p] \quad (40.22)$$

In "Two-Tier" quantum field theory we find the vacuum expectation value of a free field is zero *and the expectation value of the square of the field is also zero:*

---

[122] We can arrange for $z_i^0 = 0$ by making a Complex Lorentz transformation to an inertial frame where z is real.

$$<0|\phi(X)\phi(X)|0> = \int d^3p\, e^{-p^ip^j\Delta_{Tij}(0)/Mc^4}/[(2\pi)^3 2\omega_p] = 0$$

since the exponential factor in the integral is $-\infty$. The exponent contains

$$\Delta_{Tij}(z) = \int d^3k\, e^{-ik.z}\,(\delta_{ij} - k_ik_j/\mathbf{k}^2)/[(2\pi)^3 2\omega_k] \qquad (40.23)$$

where "T" is for "Two-Tier". Thus *vacuum fluctuations are zero in Two-Tier quantum field theory*. Correspondingly, we will see that renormalization constants are finite in the Two-Tier versions of QED, Electroweak Theory, the Standard Model and Quantum Gravity. See Blaha (2017b) and references therein for more details.

40.10 Time Intervals in General Relativity

Wigner[123] has studied the measurement of time intervals in General Relativity and sees a problem in the measurement of extremely short intervals. According to Wigner, the measurement of a time interval in a region of space requires the measurement of the length of time required for an event to happen. The measurement requires an accurate clock. But the accuracy of the clock is limited by the energy-time uncertainty relation:

$$\Delta E \Delta t \geq \hbar \qquad (40.24)$$

Thus the uncertainty in the clock's time measurement is related to the uncertainty in the clock's energy which is, in turn, related to the uncertainty in the clock's mass:

$$\Delta E = (\Delta m)c^2$$

To obtain "infinite" accuracy the uncertainty (fluctuations) in the clock's mass must be infinite and thus the clock's mass must be infinite. Infinite fluctuations in the clock's mass will produce corresponding infinite fluctuations in the gravitational field.

$$\Delta h \propto \Delta E \qquad \text{(in conventional General Relativity)}$$

As a result the notions of space-time and time intervals (which depend on the geometry through General Relativity) become uncertain. Thus, according to Wigner, and others, the concept of time intervals and space-time points becomes questionable.

The Two-Tier version of Quantum Gravity offers a way out of this dilemma. The gravitational force becomes stronger as one goes to shorter distances (higher energies) down to a distance (or up to an energy) whose scale is set by M_c. At shorter distances (higher energies) the gravitational force becomes weaker and declines to zero at zero distance. Thus at very high energy the gravitational field fluctuations (Δh) are at worst inversely proportional to the energy (and probably decline by a higher power of

[123] E. P. Wigner, Rev. Mod. Phys. **29**, 255 (1957); J. Math. Phys. **2**, 207 (1961).

inverse energy.) (The same considerations would apply if one chooses to consider fluctuations in the Riemann-Christoffel symbols.)

$$\Delta h < c_1/E < c_1/(\Delta E) \quad \text{(in Two-Tier Quantum Gravity)} \qquad (40.25)$$

where c_1 is a constant. Thus Wigner's conclusion does not hold in the Two-Tier version of Quantum Gravity as gravitational fluctuations actually become smaller at energies above a critical energy whose scale is set by M_c.

In fact, combining the above equations we see

$$c_1\Delta t/\Delta h \geq \hbar$$

at sufficiently high energy. Therefore the time uncertainty Δt, and the gravitational field fluctuations Δh, can both decrease while maintaining the energy-time uncertainty relation. *Thus the notion of a space-time point "is saved" in Two-Tier quantum gravity.*

40.11 Vacuum Fluctuations in the Gravitation Fields

While the expectation value of the free graviton field $h_{\mu\nu\text{conv}}(x)$ (weak field approximation) is zero in a conventional quantum field theoretic approach:

$$<0|h_{\mu\nu\text{conv}}(x)|0> = 0 \qquad (40.26)$$

the vacuum fluctuations of the *conventional* quantum graviton field is quadratically divergent since

$$<0|h_{\mu\nu\text{conv}}(x)h_{\alpha\beta\text{conv}}(x)|0> = \int d^3p\, b'_{\mu\nu\alpha\beta}(p)/[(2\pi)^3\, 2\omega_p] = \infty \qquad (40.27)$$

where $b'_{\mu\nu\alpha\beta}(p)$ is a rational function of the momentum p.

In "Two-Tier" quantum field theory we find

$$<0|h_{\mu\nu}(X)h_{\alpha\beta}(X)|0> = \int d^3p\, b'_{\mu\nu\alpha\beta}(p)\, e^{-p^ip^j\Delta_{Tij}(0)}/[(2\pi)^3 2\omega_p] = 0 \qquad (40.28)$$

since the exponential factor in the integrand is $-\infty$. The exponent contains

$$\Delta_{Tij}(z) = \int d^3k\, e^{-ik\cdot z}(\delta_{ij} - k_ik_j/\mathbf{k}^2)/[(2\pi)^3 2\omega_k]$$

Thus the vacuum fluctuations of $h_{\mu\nu}$ are zero in "Two-Tier" quantum field theory and, correspondingly, the weak field Two-Tier quantization of Quantum Gravity is consistently finite (and weak in perturbation theory calculations.)

40.12 Two-Tier Features in D-Dimensional Space-Time (such as the Megaverse)

Since a field, quantized in D-dimensional conventional coordinates (D > 4), would lead to divergences in perturbation theory calculations, we can use D-dimensional Two-Tier coordinates to avoid divergences in perturbation theory:

$$Y^i(y) = y^i + i\, Y_u^{\,i}(y)/M_u^{\,D/2} \tag{40.29}$$

where $Y_u^{\,i}(y)$ for i = 1, …, D is a D-dimensional free gauge field and M_u is a mass of the order of the Planck mass or greater. The $Y_u^{\,i}(y)$ term adds a quantum field to the D coordinates making them a set of quantum coordinates. Quantum coordinate derivatives are defined by

$$\partial_i = \partial/\partial Y^i(y) = \partial/\partial(y^i - Y_u^{\,i}(y)/M_u^{\,D/2}) \tag{40.30}$$

The use of these coordinates to quantize particle fields leads to a completely finite perturbation theory. We applied them in Blaha (2017b) to create a finite fundamental theory of mater. We applied them to fields in the Megaverse[124] to achieve a finite theory of Megaverse dynamics for elementary particles and universe particles.

The second quantization of a vector gauge field $V^i(y)$ is analogous to the second quantization of the electromagnetic field. The lagrangian density terms for the free $V^i(Y(y))$ fields is

$$\mathscr{L}_{Vu} = -\tfrac{1}{4} F_{Vu}^{\;\;ij}(Y(y)) F_{Vuij}(Y(y)) \tag{40.31}$$

The lagrangian is

$$L_{Vu} = \int d^Dy\, \mathscr{L}_{Vu}(Y(y))$$

with

$$F_{Vuij} = \partial V_i(Y(y))/\partial Y^j(y) - \partial V_j(Y(y))/\partial Y^i(y)$$

where the values of i and j range from 1 to D in this section.

The equal time commutation relations, using the Dth coordinate as the time coordinate, are specified in the usual way:

$$[V^i(Y(\mathbf{y}, y^0)), V^j(Y(\mathbf{y}', y^0))] = [\pi^i(Y(\mathbf{y}, y^0)), \pi^j(Y(\mathbf{y}', y^0))] = 0$$
$$[\pi_j(Y(\mathbf{y}, y^0)), V_k(Y(\mathbf{y}', y^0))] = -i\,\delta^{(D-1)tr}{}_{jk}(Y(\mathbf{y},0) - Y(\mathbf{y}',0))$$

where

$$\pi_u^{\,k} = \partial\mathscr{L}_{Vu}(V(Y(y)))/\partial V_k'(Y(y))$$
$$\pi_u^{\,D} = 0$$

for k = 1, … , (D – 1), and

[124] Blaha (2017c).

$$\delta^{(D-1)tr}{}_{jk}(\mathbf{y} - \mathbf{y'}) = \int d^{(D-1)}k \; e^{i\,\mathbf{k}\cdot(\mathbf{Y}(\mathbf{y},0) - \mathbf{Y}(\mathbf{y'},0))} \, (\delta_{jk} - k_j k_k/\mathbf{k}^2)/(2\pi)^{D-1} \tag{40.32}$$
$$V_k'(Y(y)) = \partial V_k(Y(y))/\partial y^{1D}$$

for j, k = 1, 2, … , (D − 1).

If we choose the Radiation gauge for $V_k(Y(y))$:

$$V^D(Y(y)) = 0$$
$$\partial V^j(Y(y))/\partial Y^j(y) = 0 \tag{40.33}$$

for j = 1, 2, … , (D − 1) then (D − 2) degrees of freedom (polarizations) are present in the vector potential.[125] The Fourier expansion of the vector potential $V^i(Y(y))$ is:

$$V^i(Y(y)) = \int d^{(D-1)}k \; N_{0V}(k) \sum_{\lambda=1}^{D-2} \varepsilon^i(k, \lambda)[a_V(k,\lambda) :e^{-ik\cdot Y(y)}: + a_V^\dagger(k,\lambda) :e^{ik\cdot Y(y)}:] \tag{40.34}$$

for i = 1, … , (D − 2) where

$$N_{0V}(k) = [(2\pi)^{(D-1)} 2\omega_k]^{-\frac{1}{2}}$$

and (since the field is massless)

$$k^D = \omega_k = (\mathbf{k}^2)^{\frac{1}{2}}$$

where k^D is the energy, and where the $\varepsilon^i(k, \lambda)$ are the polarization unit vectors for λ = 1, … , (D − 2) and $k^\mu k_\mu = k^{D\,2} - \mathbf{k}^2 = 0$.

The commutation relations of the Fourier coefficient operators are:

$$[a_V(k,\lambda), a_V^\dagger(k',\lambda')] = \delta_{\lambda\lambda'}\delta^{D-1}(\mathbf{k} - \mathbf{k'})$$
$$[a_V^\dagger(k,\lambda), a_V^\dagger(k',\lambda')] = [a_V(k,\lambda), a_V(k',\lambda')] = 0$$

and the polarization vectors satisfy

$$\sum_{\lambda=1}^{D-2} \varepsilon_i(k, \lambda)\varepsilon_j(k, \lambda) = (\delta_{ij} - k_i k_j/\mathbf{k}^2)$$

The V^μ Feynman propagator is

$$iD_F^{trTT}(y_1 - y_2)_{jk} = \langle 0|T(V_j(Y(y_1))V_k(Y(y_2)))|0\rangle \tag{40.35}$$

$$= -ig_{jk} \int \frac{d^D k \; e^{-ik.(y_1 - y_2)} \, R(\mathbf{k}, y_1 - y_2)}{(2\pi)^D \, (k^2 + i\varepsilon)}$$

where g_{jk} is the D-dimensional Lorentz metric and where $R(\mathbf{k}, y_1 - y_2)$ is given by

[125] Note we use the Radiation gauge for $Y^\mu(y)$ also.

$$R(\mathbf{k}, y_1 - y_2) = \exp[- k^i k^j \Delta_{Tij}(y_1 - y_2)/M_u^{D}]$$
$$= \exp\{ -k^2[A(v) + B(v)\cos^2\theta] / [(2\pi)^{D-2} M_u^4 z^2]\}$$

where k^2 is *the sum of the squares of the D – 1 spatial components* with

$$z^\mu = y_1^\mu - y_2^\mu$$
$$z = |\mathbf{z}| = |\mathbf{y_1} - \mathbf{y_2}|$$
$$k = |\mathbf{k}|$$
$$v = |z^0|/z$$
$$A(v) = (1 - v^2)^{-1} + .5v \ln[(v - 1)/(v + 1)]$$
$$B(v) = v^2(1 - v^2)^{-1} - 1.5v \ln[(v - 1)/(v + 1)]$$
$$\mathbf{k}\cdot\mathbf{z} = kz \cos\theta$$

and $|\mathbf{k}|$ denoting the length of a spatial (D – 1)-vector **k** while $|z^0|$ is the absolute value of $z^0 \equiv z^D$.

As the above equations indicate, the Gaussian damping factor R(k, z) for *all* large spatial momentum k^j is the same for both the positive and negative frequency parts of the (Two Tier) V Feynman propagator. We are assuming the spatial momentum is real-valued in this discussion. It is also important to note that R(k, z) does not depend on $k^0 = k^D$ (in the V and Y_u Radiation gauges) and thus the integration over k^0 proceeds in the usual way to produce time-ordered positive and negative frequency parts.

The Gaussian exponential factor in *all* spatial coordinates causes the Feynman propagator to be finite and, together with the Gaussian factor in universe particle propagators, causes all perturbation theory calculations when interactions are introduced to be finite as we have seen in Blaha (2017b).

For small momentum much less than M_u then $R(\mathbf{k}, y_1 - y_2) \rightarrow 1$ and the Feynman propagator is the "normal" propagator of conventional D-dimensional quantum field theory. For large momentum the corresponding potential approaches r^{D-3} in contrast to the electromagnetic Coulomb potential r^{-1}. The V potential is highly non-singular at large energies.

Thus using Two-Tier Quantum Field Theory we can perform perturbation theory calculations that always yield a finite result.[126] This is not true if conventional Quantum Field is used.[127]

---

[126] In particular, the fermion triangle divergence (anomaly) does not occur in our Two Tier Quantum Field Theory of the fermion sector. Thus there is no requirement for axion-like particles in the Megaverse (or in universes) although the possible existence of this type of particle is not ruled out.

[127] Blaha (2005a) provides a complete discussion of Two-Tier Quantum Field Theory.

41. PseudoQuantum Field Theory

PseudoQuantum Field Theory (and its Quantum Mechanics analogue CQ Mechanics[128]) originates in the need to second quantize in unusual coordinate systems, and in curved space-time coordinate systems. The paper in Appendix 1-A provide a detailed introduction to PseudoQuantum Field Theory.

In this subsection we point out its advantages in a variety of field theory contexts that are relevant for the Unified SuperStandard Theory. The advantages of PseudoQuantum Field Theory are:

1. Quantization in any coordinate system in flat or curved space-times with an invariant definition of asymptotic particle states. An n particle asymptotic state in one coordinate system is a unitarily equivalent n particle asymptotic state in any other coordinate system. Therefore particle number is invariant under change of coordinate system. This is important for the Unified SuperStandard Theory in curved space-times. It is also important for quantization in higher dimensional Euclidean spaces such as the Megaverse. The method was developed in the late 1970's by the author to provide a quantization procedure which supports a unique particle interpretation of states in arbitrary non-static space-times where no global timelike coordinate (Killing vector) exists. PseudoQuantum Field Theory which we developed in a series of books[129] also can be formulated in the Megaverse. Thus we can use it in the Megaverse to implement the Higgs Mechanism to generate particle masses and symmetry breaking.

2. PseudoQuantum Field Theory enables one to define Higgs particle dynamics in such a way that a non-zero vacuum expectation value cleanly separates from the quantum field part of the Higgs fields. This technique can be used in symmetry breaking mechanisms, mass generation, and possible generation of coupling constants as vacuum expectation values.

---

[128] See Blaha (2016f) for CQ Mechanics, which encompasses both classical mechanics and quantum mechanics, and provides a method of rotating between them. It has applications to transitions between Quantum/Semi-Classical Entanglement, and Quantum/Classical Path Integrals, and Quantum/Classical Chaos.

[129] See Blaha (2017b) for the discussion of the PseudoQuantum field theory formalism for Higgs particles in our Extended Standard Model. See chapter 20 of Blaha (2017b), and earlier books, for a more detailed view than that presented here.

3. It supports the canonical definition of higher derivative field theories through the use of the Ostrogradski bootstrap. See Appendix B where a fourth order theory of the Strong interaction is defined that has color confinement and a linear r potential. The potential part of this theory was used by the Cornell group to calculate the Charmonium spectrum. (See Blaha (2017b) for details.)

An associated advantage of using PseudoQuantum Field Theory is that it provides for retarded propagators and an Arrow of Time.

41.1 General Case of PseudoQuantization in Differing Coordinate Systems

Papers in Appendix A describe the PseudoQuantization procedure that relates second quantizations in differing coordinate systems. We can epitomize the general concept in the following short example.

Consider the case of a scalar particle in D space-time dimensions that we second quantize in coordinate system denoted 1 with coordinates x based on a timelike Killing vector

$$\varphi(x) = \sum_{\alpha} [\chi_\alpha(x)A_\alpha + \chi_\alpha{}^*(x)A_\alpha{}^\dagger] \tag{41.1}$$

where the $\chi_\alpha(x)$ are positive frequency with respect to a definition of positive frequency within a universe – following the notation of Appendix A.

Consider now the second quantization of the particle field in a second coordinate system denoted 2 with coordinates y based on a different timelike Megaverse Killing vector

$$\varphi(y) = \sum_{\beta} [\psi_\beta(y)b_\beta + \psi_\beta{}^*(y)b_\beta{}^\dagger] \tag{41.2}$$

where the $\psi_\beta(y)$ are positive frequency with respect to 2's definition of positive frequency.

Comparing above definitions we see the difference in the definition of the coordinates used in the field expansions as well as the implicit difference in the definitions of positive frequency. To relate the quantizations to each other, we must use the relation between the x and y coordinates:

$$y_i = f_i(x)$$

or, in vector form,

$$y = f(x)$$

for i = 1, 2, … , D. Thus

$$\varphi(f(x)) = \sum_{\beta} [\psi_\beta(f(x))b_\beta + \psi_\beta{}^*(f(x))b_\beta{}^\dagger] \tag{41.3}$$

Inverting the above equations to obtain the relation of the fourier coefficient operators we see:

$$A_\alpha = \sum_\beta [C_{\alpha\beta}\, b_\beta + C'_{\alpha\beta}\, b_\beta^\dagger]$$

where $C_{\alpha\beta}$ and $C'_{\alpha\beta}$ are c-number functions of α and β:

$$C_{\alpha\beta} = (\chi_\alpha(x), \varphi(f(x))) \qquad (41.4)$$
$$C'_{\alpha\beta} = (\chi_\alpha{}^*(x), \varphi(f(x))$$

The above equations imply an N particle state in one coordinate system will appear as a superposition of states of various numbers of particles in the other coordinate system IF the standard quantum field theory formulation is used.

TO REMEDY this situation – which we take to be unphysical – we must reformulate quantum field theory using the PseudoQuantum formulation presented Appendix 1-A. The scalar particle case is discussed in Appendix 1-A.

The conclusions of that section, and the sections following it in Appendix 1-A, are:

1. One can define corresponding unitarily equivalent particle states in two quantizations with invariant particle numbers.

2. The fourier coefficient operators of the two quantizations are related by Bogoliubov transformations and are unitarily equivalent.

3. The group of the local Bogoliubov transformations is an infinite tensor product of $SU_{1,1}$ groups.

4. The vacuums of the particles are invariant under Bogoliubov transformations that relate the Megaverse and the universe quantizations.

5. Unitarily equivalent perturbation theories of both quantizations can be defined.

We now consider the case of Two-Tier PseudoQuantization, and then turn to various applications of PseudoQuantization.

41.2 Two-Tier PseudoQuantum Field Theory

The combination of the Two-Tier procedure with the PseudoQuantization procedure leads to a somewhat more complicated situation. In principle, both are required for a Unified SuperStandard Theory in any coordinate system in flat or curved space-times in any number of dimensions. However their direct combination is both complicated and unphysical.

The main purpose of PseudoQuantization is to have particle number invariance under a change of coordinate system. Two-Tier Field Theory ‘cloaks’ each particle in infinite ‘clouds’ of Y^μ quanta as Fig. 40.1 illustrates. We define PseudoQuantization as implementing particle number invariance for ‘bare’ particles without their clouds of Y^μ quanta. Thus an asymptotic particle state of n particles (neglecting its Y^μ quanta cloud) remains a unitarily equivalent n particle state (neglecting its Y^μ quanta cloud) under a change of coordinate system.

To implement this concept we first define quantizations of a particle in coordinate systems without Two-tier quanta. We then 'dress' the quantizations by replacing the coordinates y^μ in each coordinate system with the corresponding Two-Tier coordinates:

$$y^\mu \rightarrow X^\mu(y) = y^\mu + i\, Y^\mu(y)/M_c^2 \qquad (41.5)$$

It appears the most convenient gauge in each coordinate system is the Lorentz gauge:

$$\partial Y^\mu/\partial y^\mu = 0 \qquad (41.6)$$

We now briefly consider the case of a scalar particle PseudoQuantization. We must introduce two fields $\varphi_1(y)$ and $\varphi_2(y)$ with the free fields' lagrangian

$$\mathscr{L}(y) = \partial^\mu\varphi_1\partial_\mu\varphi_2 - \tfrac{1}{2}\,\partial^\mu\varphi_1\partial_\mu\varphi_1 - m^2\varphi_1\varphi_2 + \tfrac{1}{2}\,m^2\,\varphi_1^{\,2} \qquad (41.7)$$

in a coordinate system with coordinates y. Then we arrive at a PseudoQuantum formulation in the coordinate system with coordinates y that is unitarily equivalent to that of a different coordinate system defined a similar manner.

We can replace the c-number coordinates x and y with Two-Tier coordinates of the form

$$X^\mu(y) = y^\mu + i\, Y^\mu(y)/M_c^2$$

and proceed to calculate propagators and perturbation theory diagrams.

Thus we have a straight-forward procedure to unite the PseudoQuantum formalism with Two-Tier coordinates to obtain finite perturbation theory results with unitary equivalence to quantization in other coordinate systems in both flat and curved space-times.

The use of two fields per particle of PseudoQuantum field theory will be seen to part of the applications consider in the remainder of this subsection. We will put aside the consideration of quantizations in other coordinate systems in what follows to keep the presentation as simple as possible.

41.3 PseudoQuantum Higgs Scalar Particle Field Theory in D-dimensional Space-Time

41.3.1 The Enigma of Higgs Particles and the Higgs Mechanism

In our previous work on the Standard Model, and its generalization to The Unified SuperStandard Theory described in a series of books entitled *Physics is Logic ...*, we showed that the fermion spectrum results from Complex Special Relativity, the gauge interactions result from the Reality group, the fermion generations result from the Generation group, and the Theory of Everything results from a combination with

Complex General Relativity. The Higgs particles and the Higgs Mechanism were inserted to generate particle masses and symmetry breaking effects.

Whence come Higgs particles? A more fundamental cause has not been suggested until our analysis, which is presented here. So the Higgs sector appeared to be an expedient mechanism to insert much needed symmetry breaking and masses into the theory.

There are a number of peculiarities in the implementation of the Higgs Mechanism:

1. First, it is selective in the sense that some gauge fields have associated Higgs particles and utilize the Higgs Mechanism, and some gauge fields do not have associated Higgs particles. In particular, the ElectroWeak gauge fields, the Generation group gauge fields, the Layer group fields, and the complex gravity Species gauge fields have associated Higgs particles. The strong interaction (gluon) gauge fields do not.

2. The Higgs potentials have a quadratic mass term of the "wrong" sign plus a quartic interaction term, which together, generate non-zero vacuum expectation values. They obviously accomplish their goal. But the source of these potentials, and why they have the same form, is unknown. One expects a fundamental principle should be operative here.

3. One can imagine creating a Higgs microscope at some super-accelerator. Using this microscope in the presence of a (classical) condensate could enable the Uncertainty Principle to be violated. This possibility, in the case of a microscope using electromagnetic fields, was the source of a heuristic argument for the need to quantize the electromagnetic field.[130]

4. The formulation of the Higgs Mechanism uses classical fields under the assumption that a path integral formulation justifies their use. While this may be true, the path integral formulation relies on implicit, unstated boundary conditions that obscure the physics of the quantum field theoretic nature of the mechanism. A direct quantum field theoretic study of the Higgs Mechanism is needed and would further elucidate its character.

5. Scalar fields have a cloud hanging over them that spin ½ fields do not. A spin ½ particle cannot transition to negative energy because there is a filled sea of negative energy particles. No additional particles can fall into the sea due to the Pauli Exclusion Principle that forbids two fermions with the same 4-momentum and quantum numbers. In the case of scalar particles the Pauli Exclusion Principle does not apply and so a *filled* negative energy sea of

[130] Heitler (1954) p. 86 provides a good discussion of the need to quantize the electromagnetic field.

scalar particles is not possible and positive energy scalar particles can transition to negative energy without hindrance. This problem has been "resolved" by an appropriate definition of the scalar particle vacuum to exclude transitions to negative energy. But the rationale for the definition is lacking. Dirac was asked about this issue many years ago. He said he had a solution to the problem. However he did not present it – in keeping with his well-known taciturn nature. So the issue remains an open question.

For the above reasons we will show that a more satisfactory method of achieving the goals of mass generation and symmetry breaking exists.[131] This method relies on a larger Fock space that enables the appearance of a vacuum expectation value for Higgs particles to be understood within a truly quantum framework. More importantly, this method is a consequence the PseudoQuantization procedure described above that enables unitarily equivalent quantizations in different coordinate systems. So a profound fundamental justification for our Higgs boson formulation exists. One major consequence of this approach is the appearance of a local Arrow of Time – a concept that has been a subject of interest for over one hundred years. Another consequence is a rationale for ElectroWeak Higgs bosons and for their absence for the strong (gluon) interaction.

41.3.2 PseudoQuantization of Scalar Particles

We now consider the PseudoQuantization[132] of a scalar particle field that will become a Higgs particle with a non-zero vacuum expectation value.[133] We begin by defining two fields that correspond to the scalar particle: $\varphi_1(x)$ and $\varphi_2(x)$.[134] These fields will be assumed to have the equal time commutators

$$
\begin{aligned}
&[\varphi_i(x), \pi_j(y)] = i(1 - \delta_{ij})\delta^3(\mathbf{x} - \mathbf{y}) \qquad (41.8)\\
&[\varphi_i(x), \varphi_j(y)] = 0\\
&[\pi_i(x), \pi_j(y)] = 0
\end{aligned}
$$

where δ_{ij} is the Kronecker δ and where $\pi_i(x)$ is the canonically conjugate momentum to $\varphi_i(x)$. The fields $\varphi_1(x)$ and $\pi_1(y)$ will be observable classical fields. The fields $\varphi_2(x)$ and

---

[131] In the Extended Standard Model of Blaha (2015a) we have shown that the basic particles have a mass, the Landauer mass, so that the theory is symmetry violating from the very start. We have also shown that our Two-Tier formalism for quantum field theories always yields finite results in perturbation theory calculations – making the renormalization approach of t'Hooft and others, which relied on initially massless gauge fields, unnecessary.

[132] PseudoQuantization in a D-dimensional space-time is described in Blaha (2017c). This discussion is relevant to PseudoQuantization in the Megaverse, or in other universes.

[133] Much of this section appears in Blaha (2016c), and earlier books, as well as in S. Blaha, Phys. Rev. **D17**, 994 (1978). The case of fermion PseudoQuantization is also discussed in Appendix A – S. Blaha, Il Nuovo Cimento **49A**, 35 (1979).

[134] The subscripts on the fields are not gauge symmetry indices but simply identifiers distinguishing the fields from each other.

$\pi_2(y)$ will not be observables so that $\varphi_1(x)$ and $\pi_1(y)$ can both be sharp on the set of physical states.

We now specify the lagrangian density for a scalar Klein-Gordon particle:

$$\mathcal{L} = \partial\varphi_1/\partial x_\mu \partial\varphi_2/\partial x^\mu \tag{41.9}$$

with hamiltonian density

$$\mathcal{H} = \pi_1\,\pi_2 + \partial\varphi_1/\partial x_i \partial\varphi_2/\partial x^i$$

where i labels spatial coordinates, and $\pi_1 = \partial\varphi_2/\partial t$ and $\pi_2 = \partial\varphi_1/\partial t$. The lagrangian $\mathcal{L}$ is without a potential or mass term.

The lagrangian and hamiltonian for a massive scalar particle in this formalism are

$$\mathcal{L} = \partial\varphi_1/\partial x_\mu \partial\varphi_2/\partial x^\mu - m^2\,\varphi_1\varphi_2 \tag{41.10}$$

with hamiltonian density

$$\mathcal{H} = \pi_1\,\pi_2 + \partial\varphi_1/\partial x_i \partial\varphi_2/\partial x^i + m^2\,\varphi_1\varphi_2$$

The fields can be fourier expanded in terms of creation and annihilation operators:

$$\varphi_i(\mathbf{x}, t) = \int d^3k\,[a_i(k)f_k(x) + a_i^\dagger(k)f_k{}^*(x)] \tag{41.11}$$

for i = 1, 2 where

$$f_k(x) = e^{-ik\cdot x}/(2\omega_k(2\pi)^3)^{1/2}$$

with $\omega_k = |\mathbf{k}|$.

The creation and annihilation operators satisfy the commutation relations:

$$[a_i(k), a_j^\dagger(k')] = (1 - \delta_{ij})\delta^3(\mathbf{k} - \mathbf{k'})$$
$$[a_i(k), a_j(k')] = 0$$
$$[a_i^\dagger(k), a_j^\dagger(k')] = 0$$

for i, j = 1, 2.

In this formulation the defining properties of a physical state are:

$$\varphi_1(x)|\Phi, \Pi> = \Phi(x)|\Phi, \Pi> \tag{41.12}$$
$$\pi_1(x)|\Phi, \Pi> = \Pi(x)|\Phi, \Pi>$$

where $\Phi(x)$ and $\Pi(x)$ are sharp on the states and thus classical fields with

$$\Phi(\mathbf{x}, t) = \int d^3k\,[\alpha(k)f_k(x) + \alpha^*(k)f_k{}^*(x)] \tag{41.13}$$

and correspondingly for $\Pi(x)$.

41.3.3 Vacuum States for Scalar (Higgs) Particles with Non-Zero Vacuum Expectation Values

When we implement the mass mechanism, Φ is constant. We can define a set of states

$$a_1(k)|\alpha> = \alpha(k)|\alpha>$$
$$a_1^\dagger(k)|\alpha> = \alpha^*(k)|\alpha>$$

and correspondingly a set of coherent states

$$|\alpha> = C\exp\{\int d^3k\ [\alpha(k)a_2^\dagger(k) + \alpha^*(k)a_2(k)]\}|0> \qquad (41.14)$$

where C is a normalization constant and where the vacuum state |0> satisfies

$$a_1(k)|0> = a_1^\dagger(k)|0> = 0 \qquad (41.15)$$

$$a_2(k)|0> \neq 0 \qquad\qquad a_2^\dagger(k)|0> \neq 0$$

The dual vacuum state satisfies

$$<0|a_2(k) = <0|a_2^\dagger(k) = 0$$

$$<0|a_1(k) \neq 0 \qquad\qquad <0|a_1^\dagger(k) \neq 0$$

With this coherent state formalism, which gives purely classical fields and yet also has quantum fields through the use of φ_2 and its creation and annihilation operators, we now have the machinery to define a mass mechanism without the introduction of a potential whose origin can only be described as dubious.

For we can define a coherent state for some k as

$$|\Phi, \Pi> = C\exp\{[(2\pi)^3\omega_k/2]^{1/2}\Phi[a_2^\dagger(k) + a_2(k)]\}|0> \qquad (41.16)$$

where C is a normalization constant, which yields a non-zero vacuum expectation value:

$$\varphi_1(x)|\Phi, \Pi> = \Phi|\ \Phi, \Pi> \qquad (41.17)$$

where Φ is a constant. Evaluating a fermion interaction term we find a mass term emerges[135]

$$\psi\ (\varphi_1 + \varphi_2)\psi\ \rightarrow \bar{\psi}(\Phi + \varphi_2)\psi \qquad (41.18)$$

It generates a mass for an interaction with a gauge field of the form

$$A^\mu(\varphi_1 + \varphi_2)^2A_\mu\ \rightarrow A^\mu(\Phi + \varphi_2)^2A_\mu \qquad (41.19)$$

It also yields a quantum field theoretic interaction that would result in the production of ElectroWeak particles from these scalar fields. The production of Higgs particles that decay into ElectroWeak gauge particles has recently been found at CERN.

---

[135] When matrix elements with a "vacuum state" are taken.

The present formalism provides a clean way to separate the vacuum expectation value of a scalar particle from its quantum field part in contrast to the Higgs Mechanism where one has to separate a Higgs field into parts manually.

41.3.4 Interpretation of Negative Energy Scalar Particle States

As we noted earlier, scalar particle physics has the problem of no barrier to the decay of positive energy states to negative energy states due to the absence of a Pauli Exclusion Principle for bosons. The PseudoQuantization procedure that we developed in 1978 and describe here allows negative energy states as one would physically expect and raises the possibility of disastrous particle decays to negative energy. The above equations show that negative energy states are possible in this theory.

However they also show that combined positive and negative energy boson states can be interpreted as classical field states. In addition, the ability of any number of boson particles to have the same 4-momentum and quantum numbers shows that a *macroscopic* classical scalar field state can be constructed.

Thus we can view states containing negative energy particles as classical field states and thus solve[136] *the issue of interpreting negative energy particle states – a more satisfactory approach than the standard quantization procedure does – with due respect to Professor Dirac.*

We note that macroscopic many particle fermion states can only have one particle in any mode unlike bosons. Therefore we cannot use this formalism to create macroscopic classical fermion field states.[137] And the filled Dirac sea of negative energy fermions precludes the transition of a positive energy Dirac fermion to a negative energy state. *Thus there is a certain complementarity between fermions that cannot become classical fields but have a filled sea precluding decays to negative energy states, and bosons that can become classical fields but support decays to negative energy states.*

41.3.5 Contrast with Conventional Second Quantization of Scalar Particles

The PseudoQuantization procedure followed here uses different boundary conditions than the usual scalar particle quantization procedure. The essence of the difference is embodied in a comparison of the definition of the vacuum above and the definition of the conventional second quantized field vacuum:

$$a|0> = 0 \qquad \text{Conventional Approach}$$
$$a^\dagger|0> \neq 0$$

In the conventional approach the creation of negative energy boson states is eliminated *ab initio* whereas in our approach it is allowed in order to support classical field states

---

[136] Also a boson that has no interactions cannot transition from to a positive energy state to a negative energy state due to conservation of energy.

[137] However we can create PseudoQuantum fermion states. See S. Blaha, Phys. Rev. **D17**, 994 (1978) (reproduced in Appendix I) and references therein to earlier papers by the author.

with non-zero vacuum expectation values that are a form of classical field. While one cannot discredit the conventional choice for conventional scalar fields, one can see that our approach yields a physically more important result – particularly for Higgs fields – because it leads to an Arrow of Time *locally* – an important feature of physical phenomena that has been a subject of much discussion and dispute. One can say that the conventional approach sweeps the issue "under the rug" rather than seeking a deeper justification – differing from Dirac's implied notion that the issue merited attention. We will discuss the "Arrow of Time" within the framework of our PseudoQuantization approach later.

41.3.6 Why Inertial Reference Frames are Special

The great physicists of the early 20th century raised numerous questions about Special Relativity after Einstein and Poincarè's discovery. Prominent among them was the question of why inertial reference frames are of especial importance in Special Relativity, and afterwards in General Relativity.

It appears that our formulation of the mass generation mechanism sheds significant light on the reason for the special prominence of inertial frames. Earlier we considered the case of a massless PseudoQuantized scalar. We now consider massive scalars since experiments at CERN have apparently discovered a Higgs particle with a 125 GeV/c mass. The above equations describe a massive scalar particle. If the scalar is massive, then the "vacuum" state that yields a non-zero expectation value must change to

$$|\Phi, \Pi> = C\exp\{(2\pi)^3 m/2]^{1/2}[a_2^\dagger(\mathbf{0},m) + a_2(\mathbf{0},m)]\}|0> \qquad (41.20)$$

to have operators for a particle of mass m in its rest frame. Then, having established this preferred frame for a Higgs particle, in The Unified SuperStandard Theory, and requiring that invariant intervals

$$ds^2 = dt^2 - d\mathbf{x}^2 \quad \text{(in rectangular coordinates)}$$

are unchanged by a (complex or real) Lorentz transformation, we find that inertial reference frames are singled out as "special" in the sense that they are the only accessible reference frames that can be generated by a Lorentz boost/transformation from the Higgs particle rest frame. *The Higgs particle vacuum state singles out the class of inertial reference frames.*

Thus Higgs particles play a central role in establishing the basis of physical reality.

41.3.7 PseudoQuantization Reveals More Physical Consequences than the Higgs Mechanism of Scalar Particles

Earlier we pointed out that our PseudoQuantization theory of Higgs particles reveals more physical consequences than the conventional approach, which implements the Higgs Mechanism by simply using a potential term that has a minimum at a non-zero vacuum expectation value. This section shows the major results of a properly implemented mechanism. We find a better explanation of the negative energy state problem of boson field theories. We find a local arrow of time that explains the direction of time that we, and all of nature, experiences. We find the reason why inertial reference frames have a special physical significance – a result long sought by physicists.

In addition we will see in chapter 11 that real gauge fields should have an associated Higgs particle, while necessarily complex gauge fields (the Strong interaction gauge field in The Unified SuperStandard Theory) do not have an associated gauge field. These results correspond to experimental reality.

41.3.8 The T Invariance Issues of Our PseudoQuantized Scalar Particle Theory

The PseudoQuantized scalar particle hamiltonian equations are invariant under time reversal $t \rightarrow t' = -t$. The 'new' vacuum states defined above break the time reversal invariance of the theory resulting in retarded particle propagators.

The hamiltonian equations

$$[H, \varphi_1(\mathbf{x}, t)] = -i\partial\varphi_1/\partial t \qquad (41.21)$$
$$[H, \varphi_2(\mathbf{x}, t)] = -i\partial\varphi_2/\partial t$$

are invariant under time reversal. If we define a time reversal operator transformation U then the time reversed equations are

$$[UHU^{-1}, \varphi_1(\mathbf{x}, -t)] = +i\partial\varphi_1(\mathbf{x}, -t)/\partial(-t)$$
$$[UHU^{-1}, \varphi_2(\mathbf{x}, -t)] = +i\partial\varphi_2(\mathbf{x}, -t)/\partial(-t)$$

The operator U, which is unitary, transforms H into **–H**. This operation is legal because the hamiltonian – in this case – is not positive definite and admits negative energy states.[138] Thus

$$[H, \varphi_1(\mathbf{x}, -t)] = -i\partial\varphi_1(\mathbf{x}, -t)/\partial(-t)$$
$$[H, \varphi_2(\mathbf{x}, -t)] = -i\partial\varphi_2(\mathbf{x}, -t)/\partial(-t)$$

and the time reversal invariance of the equations of motion is established for this case.

Time reversal invariance is broken by our choice of vacuum states. This choice is necessary to obtain classical field states as we showed earlier. A demonstration of the time reversal symmetry breaking is presented later where we show theory has retarded propagators for particle propagation to and from asymptotic states.

Within the interaction region the particle propagators are the sum of retarded and advanced parts that combine to yield principle value propagators – not Feynman propagators. Many years ago Feynman and Wheeler championed principle value propagators for electrodynamics to obtain an action-at-a distance theory of Quantum

[138] Unlike the usual case of second quantized Klein-Gordon quantum field theory.

Electrodynamics. While their theory, and ours, differ from the standard quantum field theory approach there is no reason to view them as faulty, or having serious physical defects. The only question is whether nature chooses conventional quantum field theory or PseudoQuantized quantum field theory. In our case the need for a classical scalar particle non-zero vacuum expectation value strongly motivates our choice of pseudoquantized Higgs particles.

41.3.9 Retarded Propagators for Our Quantized Higgs Particles

In the previous section we pointed out that our PseudoQuantization Higgs theory has an arrow of time due to its boundary conditions as expressed by its definition of the vacuum state and its dual. In this section we will show that the theory uses retarded propagators for propagation to and from the interaction region to asymptotic in-states and out-states. Within an interaction region the theory uses half-retarded – half-advanced propagators. We discuss aspects of the perturbation theory and propagators of our scalar particles in this chapter.

First we note that in-states at $t = -\infty$ are composed of superpositions of $a_2(k)$ and $a_2^\dagger(k)$ creation and annihilation operators:

$$a_2(k)|0> \neq 0 \qquad\qquad a_2^\dagger(k)|0> \neq 0$$

while the out-states composed of superpositions of $a_1(k)$ and $a_1^\dagger(k)$ creation and annihilation operators:

$$<0|a_1(k) \neq 0 \qquad\qquad <0|a_1^\dagger(k) \neq 0$$

Consequently when in-state particles (x_1) propagate into the interaction region (x_2) the relevant propagators are retarded propagators with the form

$$\begin{aligned} G_{in}(x_2, x_1) &= <0|T(\varphi_{1\,in}(x_2), \varphi_{2\,in}(x_1))|0> \\ &= \theta(x_{20} - x_{10})<0|[\varphi_{1\,in}(x_2), \varphi_{2\,in}(x_1)]\,|0> \end{aligned} \tag{41.22}$$

This is a manifestly retarded propagator. The choice of vacuums clearly results in a time asymmetry giving a retarded propagation reflecting the familiar Arrow of Time.

A similar situation prevails for propagation to out-states (x_3) from the interaction (x_2) region:

$$\begin{aligned} G_{out}(x_3, x_2) &= <0|T(\varphi_{1\,out}(x_3), \varphi_{2\,out}(x_2))|0> \\ &= \theta(x_{30} - x_{20})<0|[\varphi_{1\,out}(x_3), \varphi_{2\,out}(x_2)]\,|0> \end{aligned} \tag{41.23}$$

Within the interaction region the Higgs particles have principle value propagators.

Thus we find PseudoQuantized Higgs particles embody a local Arrow of Time. The locality of the Arrow of Time is embodied in all the particles that interact with the Higgs particle. Since the mass of *every* particle – bosons and fermions – has a Higgs contribution, and thus *every* particle interacts with the Higgs particles, the Arrow of

Time permeates The Unified SuperStandard Theory as well as the more familiar Standard Model known from experiment.

41.3.10 The Local Arrow of Time

In the *Physics is Logic* series of monographs we saw that complex coordinates led to the form of the fermion spectrum, that the mapping of complex coordinates to real-valued coordinates yielded the Reality group and The Unified SuperStandard Theory gauge interactions, that Complex General Relativity led to Higgs particles that were directly united with elementary particle masses and gave us the equality of inertial mass and gravitational mass. Later we will see the reduction of complex gauge fields to real gauge fields explains the appearance of Higgs fields in The Unified SuperStandard Theory.

The PseudoQuantization procedure leads to retarded Higgs field propagators and thence to a *local* arrow of time. Many arguments have been put forward over the past hundred plus years for the Arrow of Time. Many arguments based on Statistical Mechanics, Entropy, and Boltzmann's statistical atomic theory have suggested the Arrow of Time is a global statistical consequence. This view seems to contradict the results of elementary particle experiments where a *local* Arrow of Time is evident.

Our rationale for the Arrow of Time begins with retarded Higgs fields. Then we note that Higgs field quantum interactions appear for all fermions and gauge particles. Thus all particle interactions are imbued with an Arrow of Time. Particles united to form macroscopic matter inherit their combined Arrows of Time producing the global Arrow of Time we experience.

Thus our PseudoQuantization approach offers a more satisfactory solution of the origin of the Arrow of Time.

It is remarkable that complex quantities – coordinates and fields – through the Higgs phenomena that we have considered, lead to the equality of inertial mass and gravitational mass, and an Arrow of Time. This unity of mass and time phenomena may reflect the deeper fact that we can have no practical Arrow of Time if all particles were massless, for particle dynamics at light speed would then be pointless. This view has been expressed by DeWitt, Unruh, and others who have pointed out that, physically, time is meaningful and measurable only if masses exist; the larger the mass, the more accurate the time measurement in principle.[139]

41.3.11 Space-Time Dependent Particle Masses

It is possible that the ultimate Unified SuperStandard Theory has masses that evolve with time and may also be spatially varying – different values in different parts of the universe. Presently there is no decisive evidence for this possibility although astrophysical studies continue. In this section we will describe the mechanism for space-time dependent masses.

Consider a classical field (time and spatially varying):

---

[139] No mass, no clock; no clock, no physical time. See Blaha (2015a) pp. 368-371 for a discussion including comments by DeWitt and Unruh.

$$\Phi(\mathbf{x}, t) = \int d^3k\ [\alpha(k)f_k(x) + \alpha^*(k)f_k{}^*(x)] \qquad (41.24)$$

If we define the coherent vacuum state

$$|\alpha\rangle = C \exp\{\int d^3k\ [\alpha(k)a_2{}^\dagger(k) + \alpha^*(k)a_2(k)]\}|0\rangle \qquad (41.25)$$

then

$$\varphi_1(x)|\Phi, \Pi\rangle = \Phi(x)|\Phi, \Pi\rangle$$
$$\pi_1(x)|\Phi, \Pi\rangle = \Pi(x)|\Phi, \Pi\rangle$$

where

$$\varphi_i(\mathbf{x}, t) = \int d^3k\ [a_i(k)f_k(x) + a_i{}^\dagger(k)f_k{}^*(x)] \qquad (41.26)$$

for i = 1, 2 and where

$$f_k(x) = e^{-ik\bullet x}/(2\omega_k(2\pi)^3)^{½}$$

with ω_k equal to the energy.

41.3.12 Inertial Mass Equals Gravitational Mass

From the days of Newton through Einstein[140] to the present the equality of gravitational mass and inertial mass has been a topic of interest. Mach, who played an important role, in this ongoing discussion, thought distant masses in the universe were the source of the equality. However the origin of the equality, which has been shown experimentally to very high accuracy, remained uncertain until the *Physics is Logic* series of books, in which we showed the interconnection of the Unified SuperStandard Theory and Complex Gravitation via Higgs generated masses that united gravitational and inertial mass.

In Blaha (2016h) we showed that a Complex General Relativity transformation can be factored into the product of a complex-valued transformation and a real-valued General Coordinate transformation. The set of complex valued transformations form a U(4) group that we called the General Coordinate Reality group. Later we will define the Internal Symmetry Species Group as the corresponding analogue. The Species Group has gauge fields that undergo spontaneous symmetry breaking and generate contributions to all fermion masses.

Since fermion field masses are now sums of ElectroWeak Higgs contributions, Generation group Higgs contributions, Layer group Higgs contributions, and Species group contributions, and since the gravitational Higgs fields appear in all fermion masses, the equality of inertial and gravitational mass is proven. The gravitational Higgs particles' equations depend, in part, on the gravitational field by Blaha (2016h) and so set the mass scale of gravitational mass, and thereby of all Higgs mass contributions. They set the scale of inertial masses equal to the scale of gravitational

---

[140] For example, Einstein and Grossman in 1913 stated, "The theory herein described originates in the conviction that the proportionality between the inertial and gravitational mass of a body is an exact law of nature that must be expressed as a foundation principle of theoretical physics."

masses. ***Since an expression cannot mix mass scales, the gravitational mass scale must be the same as the inertial mass scale. Inertial Mass equals gravitational mass.***

We have established the equality of inertial and gravitational mass at the short distance quantum level. In our view, this explanation is far more satisfying than basing the equality on a combination of large distance phenomena and quantum phenomena. As Einstein and Weyl have pointed out, all fundamental physics phenomena should be based on a local theory. Complex Gravity as we have constructed it, combined with the Unified SuperStandard Theory, furnishes a completely local basic Theory of Everything.

The equation above contains a coherent state $|\alpha>$ for a time and spatially varying mass. The above equations can be generalized to the case of multiple space-time varying masses.[141]

$$|\Phi_1,\Phi_2, \ldots ,\Phi_n;\Pi_1,\Pi_2, \ldots ,\Pi_n> = C \prod_{i=1}^{n} \exp\left\{\int d^3k\, [\alpha_i(k)a_{2i}^\dagger(k) + \alpha_i^*(k)a_{2i}(k)]\right\}|0> \quad (41.27)$$

Then all n mass vacuum expectation values are space-time dependent:

$$\varphi_{1i}(x)\,|\,\Phi_1, \Phi_2, \ldots , \Phi_n; \Pi_1, \Pi_2, \ldots , \Pi_n> = \Phi_i(x)\,|\,\Phi_1, \Phi_2, \ldots , \Phi_n; \Pi_1, \Pi_2, \ldots , \Pi_n> \quad (41.28)$$

Thus our formalism can accommodate space-time varying masses should they be found in the Cosmos.

41.3.13 Benefits of the PseudoQuantization Method

In this book, and in earlier work, we showed that a more physically satisfactory method for avoiding the negative energy state problem exists. This method relies on the use of a larger Fock space in which negative energy states (or partially negative energy states) are interpreted as states containing classical fields or a mix of classical fields and individual boson particles. This approach resolves the negative energy boson issue and provides a common framework for boson particles and classical boson fields.

One consequence of the PseudoQuantization method is that it enables the appearance of a vacuum expectation value for Higgs particles (a constant classical field) to be understood within a truly quantum framework. Another major consequence of this approach is the appearance of a *local* Arrow of Time due to the Higgs mass generation mechanism – a concept that has been a subject of interest for over one hundred years. A macroscopic arrow of time is often described as a statistical result. But our approach yields an arrow of time at the single particle level.

The conventional approach to boson field quantization sweeps these issues "under the rug" rather than seeking a deeper justification. It differs from Dirac's implied notion that the issue merited attention.

---

[141] The "vacuum" state $|0>$ also implicitly has factors for the vacuum expectation values used for fields that give masses to fermions and vector bosons as described in Blaha (2016h).

Another important consequence of the PseudoQuantization method is that it singles out inertial reference frames when applied to the case of Higgs particles.

Yet another more subtle consequence of boson PseudoQuantization is that it provides a rationale/explanation for the presence of ElectroWeak Higgs bosons, *and for their absence for the strong (gluon) interactions. The question of why there are no strong interaction Higgs bosons has not been previously considered to the best of this author's knowledge.*

41.4 Two-Tier Formulation and PseudoQuantization in the Megaverse

These subjects are considered in Blaha (2020c) in chapter 68 as well as in earlier books by the author.

Appendix B. Complex General Relativity & Dimension – Index Connection From Chapters 50 and 51 of Blaha (2020c)

The chapters in this appendix are from Blaha (2020c). Much of those chapters also appeared in Blaha (2016h) and (2017a). They show the connection between dimensions (coordinates) and indices in *four* dimension Complex General Relativity.

50. Complex General Relativity Reformulated

50.1 Tetrad (Vierbein) Formalism

The *vierbein* formalism begins with the Equivalence Principle that allows us to define an inertial coordinate system in the neighborhood of any point Z in space-time. We will use the notation $\varsigma^\alpha(Z)$ to denote the inertial coordinates at Z. We define a tetrad or vierbein as

$$v^\alpha{}_\mu(x) = (\partial\varsigma^\alpha(x)/\partial x^\mu)_{x=Z} \tag{50.1}$$

and, in a neighborhood of Z, we can invert the relation between ς and x to define an inverse

$$w^\mu{}_\alpha(x) = (\partial x^\mu(\varsigma)/\partial\varsigma^\alpha)_{x=X} \tag{50.2}$$

such that

$$\begin{aligned} w^\mu{}_\alpha(x)v^\alpha{}_\nu(x) &= \delta^\mu{}_\nu \\ w^\mu{}_\beta(x)v^\alpha{}_\mu(x) &= \delta^\alpha{}_\beta \end{aligned} \tag{50.3}$$

In real General Relativity all *tetrads* are real-valued. In Complex General Relativity a *tetrad* $v^\alpha{}_\mu(x)$ is complex-valued.

The metric at a curved space-time point X is defined in terms of *tetrads* as

$$\begin{aligned} g_{\rho\sigma}(x) &= \eta_{\alpha\beta}\, v^\alpha{}_\rho(x)v^\beta{}_\sigma(x) \\ g^{\rho\sigma}(x) &= \eta^{\alpha\beta}\, w^\rho{}_\alpha(x)w^\sigma{}_\beta(x) \end{aligned} \tag{50.4}$$

The inverse of a *tetrad* transformation can also be expressed as

$$w_\beta{}^\nu(x) = v_\beta{}^\nu(x) = \eta_{\beta\alpha}g^{\nu\mu}(x)v^\alpha{}_\mu(x)$$

Then a *tetrad* and its inverse satisfy

$$v^\alpha{}_\mu(x)v_\beta{}^\mu(x) = \delta^\alpha{}_\beta \tag{50.5}$$

and

$$v^{\alpha}{}_{\mu}(x)v_{\alpha}{}^{\nu}(x) = \delta^{\nu}{}_{\mu}$$

There are two general types of space-time transformations that can be performed on a tetrad.

1. A complex-valued (possibly real-valued) General Relativistic coordinate transformation:
$$v'^{\alpha}{}_{\mu}(x) = \partial x^{\nu}/\partial x'^{\mu}\, v^{\alpha}{}_{\nu}(x)$$

2. A complex-valued, local *Lorentzian transformation*
$$v'^{\beta}{}_{\mu}(x) = \Lambda(x)^{\beta}{}_{\alpha}\, v^{\alpha}{}_{\mu}(x)$$

where $\Lambda(x)^{\beta}{}_{\alpha}$ is an element of a subset of the local Complex Lorentz Group.

The local Lorentzian transformations $\Lambda(x)^{\beta}{}_{\alpha}$ consist of local Lorentz transformations that are real-valued, and complex-valued Lorentz transformations. Both types of transformations satisfy the orthogonality condition:

$$\eta_{\alpha\beta}\Lambda^{\alpha}{}_{\rho}(x)\Lambda^{\beta}{}_{\sigma}(x) = \eta_{\rho\sigma} \quad (50.6)$$

Thus the *tetrad* partakes of both local (position dependent) General Relativistic transformations and local Lorentzian transformations.

50.2 Complex General Relativistic Transformations

The General Relativistic Reality group interaction emerges from complex General Relativistic transformations. We can separate elements of the set of all complex General Coordinate transformations into a product of two factors: a real-valued General Coordinate transformation and a complex-valued General Coordinate transformation. The set of complex factors can be further factored into those that satisfy

$$\Lambda(\omega, \mathbf{u})^{T}G\Lambda(\omega, \mathbf{u}) = G \quad (50.7)$$

and those that do not. We then see that the set of those that do not satisfy the above equation form a curved space representation of the U(4) group under 'multiplication' of transformations.

The elements of the set of real and complex General Coordinate transformations whose flat complex space-time limit satisfy the above equation form the elements of the Complex Lorentz group.[142]

[142] It is this part of curved space-time General Relativity that becomes the flat space-time Complex Lorentz group, which leads to the SU(3)⊗SU(2)⊗U(1)⊗SU(2)⊗U(1)⊗SU(3) Standard Model group.

We thus find the set of all 4-dimensional complex, curved space General coordinate transformations can be visualized as in Fig. 50.1. The next section describes the interplay of the three parts displayed in Fig. 50.1.

50.3 Structure of Complex General Coordinate Transformations

Complex General Coordinate transformations can be uniquely factored into products of two terms, which will later be further factored into three factors. They have the form

$$\partial x''^{\nu}(x)/\partial x^{\mu} = U(x'')^{\nu}{}_{\beta}\, \partial x'^{\beta}(x)/\partial x^{\mu} \tag{50.8}$$

where

$$x''^{\nu}(x) = U(x'')^{\nu}{}_{\beta} x'^{\beta}$$
$$x'^{\mu}(x) = U^{-1\mu}{}_{b}(x'')\, x''^{b}$$

where $U(x')^{\nu}{}_{\beta}$ is complex and where $\partial x'^{\beta}(x)/\partial x^{\mu}$ is a purely real General Coordinate transformation.

We define

$$U\,(x'')^{\mu}{}_{\nu} = w^{\mu}{}_{a}(x'')[\exp(i \textstyle\sum_k g_k\Phi_k(x'')\tau_k)]^{a}{}_{b}\, v^{b}{}_{\nu}(x'') \tag{50.9}$$
$$U^{-1}(x'')^{\mu}{}_{\nu} = w^{\mu}{}_{a}(x'')[\exp(-i\textstyle\sum_k g_k\Phi_k(x'')\tau_k)]^{a}{}_{b}\, v^{b}{}_{\nu}(x'')$$

where the constants g_k are real, and Φ_k and τ_k are hermitean. The uniqueness of the factorization follows from the Reality group (and U(4)) property that any complex 4-vector can be uniquely mapped to any specified real 4-vector.

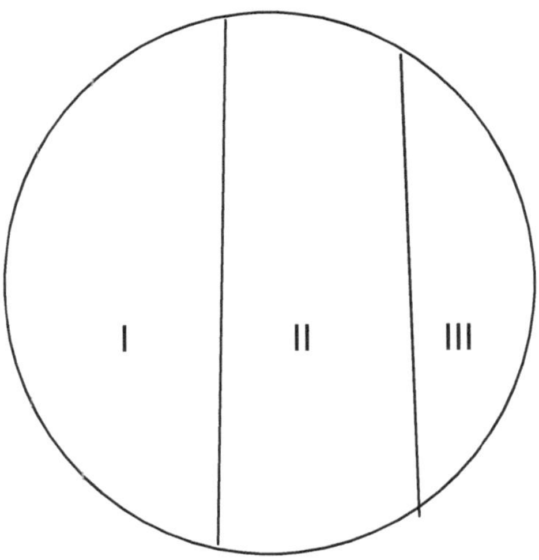

Figure 50.1. A visualization of the set of General Coordinate transformations separated into real-valued General coordinate transformations (part I), complex transformations that satisfy $\Lambda(\omega, u)^{T}G\Lambda(\omega, u) = G$ (part II), and complex transformations that do not satisfy $\Lambda(\omega, u)^{T}G\Lambda(\omega, u) = G$ (part III). Part I and

part II combine in the limit of flat space-time to form the Complex Lorentz group. Parts II and III elements form a U(4) group that we call the General Relativistic Reality group.

Given the factorization above it becomes possible to separate the affine connection correspondingly.

50.4 Complex Affine Connection – General Relativistic Reality Group

The structure of a complex general coordinate transformation enables us to calculate its affine connection for later use in determining the covariant derivative, and the dynamic equations. First the transformation to the real-valued x' coordinates from inertial coordinates is

$$\Gamma^{\sigma}{}_{\lambda\mu}(x') = \partial x'^{\sigma}/\partial\varsigma^{\rho}\ \partial^{2}\varsigma^{\rho}/\partial x'^{\lambda}\partial x'^{\mu} \tag{50.10}$$

Next the Reality group transformation has the affine connection

$$\Gamma^{\sigma}{}_{\lambda\mu}(x'') = \partial x''^{\sigma}/\partial\varsigma^{\rho}\ \partial^{2}\varsigma^{\rho}/\partial x''^{\lambda}\partial x''^{\mu}$$

which can be re-expressed as

$$\begin{aligned}\Gamma^{\sigma}{}_{\lambda\mu}(x'') &= \partial x''^{\sigma}/\partial x'^{\beta}\ \partial x'^{\beta}(\varsigma)/\partial\varsigma^{\rho}\ \partial/\partial x''^{\mu}[\partial\varsigma^{\rho}/\partial x'^{\alpha}\,\partial x'^{\alpha}/\partial x''^{\lambda}]\\ &= \partial x''^{\sigma}/\partial x'^{\beta}\ \partial x'^{\alpha}/\partial x''^{\lambda}\ \partial x'^{\gamma}/\partial x''^{\mu}\ \Gamma^{\beta}{}_{\alpha\gamma}(x') + \partial x''^{\sigma}/\partial x'^{\beta}\ \partial^{2}x'^{\beta}/\partial x''^{\lambda}\partial x''^{\mu}\end{aligned} \tag{50.11}$$

Next substituting the General Relativistic Reality group transformation

$$\begin{aligned}x''^{\nu}(x) &= U(x'')^{\nu}{}_{\beta}x'^{\beta}\\ x'^{\mu}(x) &= U^{-1}(x'')^{\mu}{}_{\beta}\,x''^{\beta}\end{aligned}$$

together with

$$\partial x''^{\sigma}/\partial x'^{\beta} = \partial[U(x'')^{\sigma}{}_{\alpha}x'^{\alpha}]/\partial x'^{\beta} = U(x'')^{\sigma}{}_{\beta} + x'^{\alpha}\,\partial U(x'')^{\sigma}{}_{\alpha}/\partial x'^{\beta}$$

$$\partial x'^{\sigma}/\partial x''^{\beta} = \partial[U^{-1}(x'')^{\sigma}{}_{\alpha}x''^{\alpha}]/\partial x''^{\beta} = U^{-1}(x'')^{\sigma}{}_{\beta} + x''^{\alpha}\,\partial U^{-1}(x'')^{\sigma}{}_{\alpha}/\partial x''^{\beta}$$

we find the second term above is the Reality fields affine connection

$$\Gamma_{R}{}^{\sigma}{}_{\lambda\mu}(x'') = \partial[U(x'')^{\sigma}{}_{\alpha}x'^{\alpha}]/\partial x'^{\beta}\ \partial\{\partial[U^{-1}(x'')^{\beta}{}_{\alpha}x''^{\alpha}]/\partial x''^{\lambda}\}/\partial x''^{\mu}$$

and so we find the affine connections are approximately additive. Thus approximately

$$\Gamma^{\sigma}{}_{\lambda\mu}(x'') = \Gamma_{GR}{}^{\sigma}{}_{\lambda\mu}(x') + \Gamma_{R}{}^{\sigma}{}_{\lambda\mu}(x'')$$

if $x''^{\sigma} \simeq x'^{\sigma}$.

A complex transformation of types II and III in Fig. 50.1 has the form:

$$U(x'')^{\mu}{}_{\nu} = w^{\mu}{}_{a}(x'')[\exp(i \sum_k \Phi_k(x'')\tau_k)]^{a}{}_{b}\, l^{b}{}_{\nu}(x'')$$
$$U^{-1}(x'')^{\mu}{}_{\nu} = w^{\mu}{}_{a}(x'')[\exp(-i\sum_k \Phi_k(x'')\tau_k)]^{a}{}_{b}\, l^{b}{}_{\nu}(x'')$$

where τ_k is a U(4) generator matrix. Its infinitesimal transformation is approximately

$$U(x'')^{\nu}{}_{\beta} \approx \delta^{\nu}{}_{\beta} + i \sum_k \Phi_k(x'')[\tau_k]^{\nu}{}_{\beta}$$
$$U^{-1}(x'')^{\nu}{}_{\beta} \approx \delta^{\nu}{}_{\beta} - i \sum_k \Phi_k(x'')[\tau_k]^{\nu}{}_{\beta} \qquad (50.12)$$

using the *vierbein* flat space-time limits

$$w^{\mu}{}_{a}(x'') \approx \delta^{\mu}{}_{a}$$
$$l^{b}{}_{\nu}(x'') \approx \delta^{b}{}_{\nu}$$

where

$$\Phi_k(x) = \int^{x} dy_{\lambda}\, A_{Rk}{}^{\lambda}(y) \qquad (50.13)$$

Then

$$\Gamma_{R}{}^{\sigma}{}_{\lambda\mu} = -\tfrac{1}{2}i\{\sum_k A_{Rk}(x'')_{\mu}[\tau_k]^{\sigma}{}_{\lambda} + \sum_k A_{Rk}(x'')_{\lambda}[\tau_k]^{\sigma}{}_{\mu}\} \qquad (50.14)$$

$$= A_{R}{}^{\sigma}{}_{\mu\lambda} + A_{R}{}^{\sigma}{}_{\lambda\mu}$$

(summed over k) with the matrix $A_{R}{}^{\sigma}{}_{\mu\lambda}$ given by

$$A_{R}{}^{\sigma}{}_{\mu\lambda} = -\tfrac{1}{2}i\sum_k A_{Rk\mu}[\tau_k]^{\sigma}{}_{\lambda} \qquad (50.15)$$

with $A_{R}{}^{\sigma}{}_{\mu\lambda}$ transformable to matrix row and column numbers

$$A_{R\,flat}{}^{\mu a}{}_{b} = A_{R\,flatk}{}^{\mu}[\tau_k]^{\sigma}{}_{\lambda}\delta_{\sigma}{}^{a}\delta^{\lambda}{}_{b}$$

using the flat space-time vierbein values, and so $A_{R\,flat}{}^{a}{}_{\mu b}$ may be written in matrix form as

$$A_{R\,flat\mu} = -\tfrac{1}{2}i\sum_k A_{R\,flatk\mu}\tau_k \qquad (50.16)$$

In the flat space-time limit the $A_{Rk}{}^{\lambda}(y)$ becomes the Coordinate Species group U(4) gauge fields $A_{R\,flatk}{}^{\lambda}(y)$.

The relevant *quadratic* $A_{R}{}^{\sigma}{}_{\mu\lambda}$ terms from eq. 50.22 below that are needed to find the dynamic equation for the gauge fields $A_{R\,flat}{}^{i}{}_{\mu}$ are contained in

$$\mathcal{L}_A = \mathrm{Tr}\, \sqrt{g}[M\partial_{\nu}R^{1}{}_{\sigma\mu}\partial^{\nu}R^{2\sigma\mu} + aR^{1}{}_{\sigma\mu}R^{2\sigma\mu} + bg^{\sigma\mu}(R^{1}{}_{\sigma\mu} + R^{2}{}_{\sigma\mu}) + 1/4(g_{\mu\nu} + g^{2}{}_{\mu\nu})T^{\mu\nu}] \qquad (50.17)$$

We can let

$$R^{i}{}_{\sigma\mu} = R^{i\beta}{}_{\sigma\beta\mu} \equiv \partial_\mu(A_R{}^{i\beta}{}_{\sigma\beta} + A_R{}^{i\beta}{}_{\beta\sigma}) - \partial_\beta(A_R{}^{i\beta}{}_{\sigma\mu} + A_R{}^{i\beta}{}_{\mu\sigma}) \tag{50.18}$$

for i = 1, 2. In the flat space-time limit we chose the Landau gauge

$$\partial_\mu A_{R\text{flat}}{}^{i\mu_a}{}_b = 0 \tag{50.19}$$

As a result

$$R^{i}{}_{\sigma\mu} \equiv \partial_\mu(A_R{}^{i\beta}{}_{\sigma\beta} + A_R{}^{i\beta}{}_{\beta\sigma}) \tag{50.20}$$

Using

$$A_{R\text{flat}}{}^{i\mu\sigma}{}_{\lambda} = A_{R\text{flat}}{}^{i\mu}[\tau_k]^a{}_b\delta^\sigma{}_a\delta_\lambda{}^b \tag{50.21}$$
$$A_{R\text{flat}}{}^{i}{}_{\mu} = -\tfrac{1}{2}i\textstyle\sum_k A_{R\text{flat}}{}^{i}{}_{k\mu}$$

and taking the trace in eq. 50.17 we obtain

$$\mathcal{L}_A = \mathrm{Tr}\,\sqrt{g}[8M\partial_\nu\partial_\mu A_{R\text{flat}}{}^{1}{}_{\sigma}\partial^\nu\partial^\mu A_{R\text{flat}}{}^{2\sigma} + 8a\partial_\mu A_{R\text{flat}}{}^{1}{}_{\sigma}\partial^\mu A_{R\text{flat}}{}^{2\sigma} + 1/4(g_{\mu\nu} + g^2{}_{\mu\nu})T^{\mu\nu}] \tag{50.22}$$

in the flat space-time limit. Eq. 50.17 needs to take account of the complex nature of $(g_{\mu\nu} + g^2{}_{\mu\nu})$ until transformed by the infinitesimal form of the complex Reality transformation:

$$\begin{aligned}(g_{\beta\alpha} + g^2{}_{\beta\alpha})' &\rightarrow U(x'')_\beta{}^\mu(g_{\mu\nu} + g^2{}_{\mu\nu})U^{-1}(x'')^\nu{}_\alpha \\ &= (\delta_\beta{}^\mu + i\textstyle\sum_k\Phi_k(x'')[\tau_k]_\beta{}^\mu)(g_{\mu\nu} + g^2{}_{\mu\nu})(\delta^\nu{}_\alpha - i\sum_k\Phi_k(x'')[\tau_k]^\nu{}_\alpha) \\ &\cong (g_{\beta\alpha} + g^2{}_{\beta\alpha}) + i\{\textstyle\sum_k \Phi_k(x'')[\tau_k]_\beta{}^\mu - i\sum_k \Phi_k(x'')[\tau_k]^\nu{}_\alpha)(g_{\mu\nu} + g^2{}_{\mu\nu}) \\ &\cong (g_{\beta\alpha} + g^2{}_{\beta\alpha}) + i\textstyle\sum_k \Phi_k(x'')\{[\tau_k]_{\beta\alpha} - [\tau_k]_{\beta\alpha}\}\end{aligned} \tag{50.23}$$

Approximating $\Phi_k(x'')$ with an infinitesimal line we find

$$\textstyle\sum_k\Phi_k(x) \cong \delta x_\lambda A_{R\text{flat}}{}^{1\lambda}(x) \tag{50.24}$$

by eq. 50.13. Thus

$$\tfrac{1}{4}(g_{\mu\nu} + g^2{}_{\mu\nu})T^{\mu\nu} \cong \tfrac{1}{4}\,[(g_{\beta\alpha} + g^2{}_{\beta\alpha}) + i\,\delta x_\lambda A_{R\text{flat}}{}^{1\lambda}(x)\{[\tau_k]_{\beta\alpha} - [\tau_k]_{\beta\alpha}\}]T^{\alpha\beta}$$

Applying the canonical Euler-Lagrange method we obtain the dynamical equations (using integration by parts to handle higher order derivative terms):[143]

$$\square^2 A_{R\text{flat}}{}^{1}{}_{\sigma} + (a/M)\square A_{R\text{flat}}{}^{1}{}_{\sigma} + i\delta x_\sigma\{[\tau_k]_{\beta\alpha} - [\tau_k]_{\beta\alpha}\}T^{\alpha\beta}/(32M) = 0 \tag{50.25}$$
$$\square^2 A_{R\text{flat}}{}^{2}{}_{\sigma} + (a/M)\square A_{R\text{flat}}{}^{2}{}_{\sigma} = 0 \tag{50.26}$$

---

[143] It is possible that the Reality transformation also depends on $A_{R\text{flat}}{}^{2}{}_{\sigma}$. Then eq. 50.26 would have an energy-momentum tensor term as well. Consequently there would be an additional interaction of the same form as in eq. 50.27 below.

Since the $A_{R_{flat}}$ gauge field is gravitational in nature it exists, as eq. 50.25 shows, as a type of gravitational interaction whose source is the energy-momentum tensor. Following the standard derivation of the gravitational potential we find the Coulomb interaction of $A_{R_{flat}}{}^{10}$.

51. Species Group U(4) Gauge Fields

From the discussion in sections 50.4 - 50.5 we see the flat space-time limit of $A_{Rk}{}^{\lambda}(y)$ is a local U(4) coordinate space gauge field. There are, *by assumption,*[144] a corresponding internal symmetry gauge fields $A_{Sk}{}^{\lambda}(y)$ – the Internal Symmetry U(4) Species Group gauge fields. The mathematical features of this field is quite similar to the U(4) Generation group fields. The interaction that appears in covariant derivatives is $g_8\mathbf{A}_S{}^{\mu}(x) = g_8\mathbf{A}_{Sk}{}^{\mu}(x)\mathbf{G}_{Sk}$ where the $\mathbf{G}_{Sk}$ are U(4) generator matrices and k is summed from 1, … , 16.

Below we will see that the effect of the Internal Symmetry Species Group is to U(4) rotate the four components of each fermion's field. Since it preserves the species of each fermion we call this group the *Species Group*. It performs a U(4) rotation of the spinor representation of each fermion

We will see that the Higgs Mechanism breakdown of the Species Group endows each fermion with a mass contribution that breaks the scale invariance of the Unified SuperStandard Theory.

Since the Species Group Higgs Mechanism breaking gives each fermion a 'gravity generated' mass, and since this mass sets the mass scale for each fermion, we conclude later that the principle of the equality of inertial and gravitational mass is a direct consequence. ***Inertial mass equals gravitational mass.***

51.1 Species Group Covariance

A Species Group transformation on a Dirac equation must be covariant. Consider the Dirac equation lagrangian term under an Internal Symmetry Species Group transformation:

$$\bar{\psi}(x)[i\gamma_\mu(\partial/\partial x_\mu - ig_8A_{Sk}{}^{\mu}(x)G_{Sk}) - m]\psi(x) = 0 \qquad (51.1)$$

summed over k. If we perform a U(4) Species group transformation U on lagrangian terms:

$$\bar{\psi}(x)[i\gamma_\mu(\partial/\partial x_\mu - ig_8A_{Sk}{}^{\mu}(x)G_{Sk}) - m]U^{-1}U\psi(x)$$

or

[144] In this discussion we *assume* that the Coordinate Species Reality Group with gauge fields $A_{Rk}{}^{\lambda}(y)$ has a corresponding Internal Symmetry Reality group that we call the Internal Symmetry Species Group. This assumption parallels the assumptions for the SU(3)⊗SU(2)⊗U(1)⊗SU(2)⊗U(1) Internal Symmetry Reality Group presented in previous chapters.

$$\bar{\psi}(x)U^{-1}U[iU^{-1}U\gamma_\mu U^{-1}U(\partial/\partial x_\mu - ig_8A_{Sk}{}^\mu(x)G_{Sk}) - m]U^{-1}U\psi(x)$$

we find

$$\bar{\psi}'(x)[i\gamma_\mu'U(\partial/\partial x_\mu - ig_8A_{Sk}{}^\mu(x)G_{Sk})U^{-1} - m]\psi'(x)$$

where

$$\gamma_\mu'(x) = U\gamma_\mu U^{-1}$$

is locally equivalent to a Dirac matrix by Good's Theorem.[145] If we set

$$A'_S{}^\mu(x) = -(i/g_8)U[\partial U^{-1}/\partial x^\mu] + UA_S{}^\mu(x)U^{-1}$$

then the transformed lagrangian terms are

$$\bar{\psi}'(x)[i\gamma_\mu'(x)(\partial/\partial x_\mu - ig_8A'_{Sk}{}^\mu(x)G_{Sk}) - m]\psi'(x) \tag{51.2}$$

They have the same form as the original terms above and thus the expression is covariant. We note the indices of the matrices G_{Sk} are spinor indices and so $G_{Sk}\gamma_\mu$ has an implicit spinor matrix summation. But the symmetry group is U(4).

The coordinate dependence of $\gamma_\mu'(x)$ introduces locality into the Dirac matrix. This locality might be viewed with concern except that an inverse Species group transformation exists that removes the locality. Thus the physical impact of this 'new' locality is eliminated.

51.2 Spontaneous Symmetry Breaking of the General Relativity U(4) Reality Group – The Species Group

We begin the discussion of the Internal Symmetry Species Group symmetry breaking[146] by defining a Higgs field η which is a Species group 4-vector

$$\eta = \begin{bmatrix} \rho_1 \\ \rho_2 \\ \rho_3 \\ \rho_4 \end{bmatrix} \tag{51.3}$$

where ρ_1, ρ_2, ρ_3 and ρ_4 are real fields.[147] Then the covariant derivative of η (taking account only of the Species group) is

---

145 R. H. Good, Jr., Rev. Mod. Phys., **27**, 187 (1955).

146 Since the Species gauge fields have been shown to have a mass it might seem redundant to introduce Higgs symmetry breaking as well. However the Higgs breaking introduces the further benefit of giving a mass term to each particle – thus establishing the equality of gravitational mass and inertial mass as we discuss in section 20.4.

147 Each field ρ_i can be expressed as a PseudoQuantum field: $\rho_i = \varphi_{1i} + \varphi_{2i}$ where φ_{1i} has the vacuum expectation value $\rho_{i0.\ \text{for}\ i = 1, \ldots, 4.}$ Thus our PseudoQuantum field theory version is implemented easily.

$$D_{\dots\mu}\eta = \{\partial/\partial X^\mu + \dots - \tfrac{1}{2}ig_8\Sigma A_{Sk}{}^\mu(x)G_{Sk}\}\begin{bmatrix}\rho_1\\ \rho_2\\ \rho_3\\ \rho_4\end{bmatrix} \quad (51.3)$$

Following steps similar to eqs. 47.4 through 47.17 for the Generation Group symmetry breaking we find with ρ_i being the vacuum expectation value of the Higgs field:

$$\begin{aligned}(D_{\dots\mu}\eta)^\dagger D_{\dots}{}^\mu\eta = {} & \partial\rho_1/\partial X^\mu\,\partial\rho_1/\partial X_\mu + \partial\rho_2/\partial X^\mu\,\partial\rho_2/\partial X_\mu + \partial\rho_3/\partial X^\mu\,\partial\rho_3/\partial X_\mu + \\ & + \partial\rho_4/\partial X^\mu\,\partial\rho_4/\partial X_\mu + \\ & + \tfrac{1}{4}\,g_8{}^2\{\rho_1{}^2A_{S1}{}^2 + \rho_2{}^2\,A_{S2}{}^2 + \rho_3{}^2A_{S3}{}^2 + \rho_4{}^2A_{S4}{}^2 + \\ & + (\rho_1{}^2 + \rho_2{}^2)(V_5{}^2 + V_6{}^2) + \tfrac{1}{4}(\rho_1{}^2 + \rho_3{}^2)(V_7{}^2 + V_8{}^2) + \\ & + (\rho_1{}^2 + \rho_4{}^2)(V_9{}^2 + V_{10}{}^2) + \tfrac{1}{4}(\rho_2{}^2 + \rho_3{}^2)(V_{11}{}^2 + V_{12}{}^2) + \\ & + (\rho_2{}^2 + \rho_4{}^2)(V_{13}{}^2 + V_{14}{}^2) + \tfrac{1}{4}(\rho_3{}^2 + \rho_4{}^2)(V_{15}{}^2 + V_{16}{}^2)\}\end{aligned} \quad (51.4)$$

up to total divergences, which generate surface terms which we discard. We also assume that all fields satisfy the gauge condition

$$\partial A_{Si}{}^\mu/\partial X^\mu = 0 \quad (51.5)$$

Eq. 51.4 shows all Species Group gauge fields have masses. Thus Species Group symmetry is completely broken. The combination of an ultra-weak coupling constant and very large gauge field masses results in extremely Species interactions.

We assume Species group gauge field masses to be very large – of the order of the Planck mass in view of its origin in Complex General Relativity.

51.3 Species Group Higgs Mechanism Contributions to Fermion Masses

The symmetry breaking of the Species Group results in a contribution to each fermion mass of all types, species, generations, and layers. The Species Group contributions to normal and Dark fermion mass terms are

$$\mathscr{L}^{Higgs}{}_{FermionMassesSpecies} = \Sigma_{s,g,l}\bar{\psi}_{sglL}\rho_s\, m_{sgl}\psi_{sglR} + \Sigma_{s,g,l}\bar{\psi}_{DsglL}\rho_s\, m_{Dsgl}\psi_{DsglR} + \text{c.c.} \quad (51.6)$$

The η field expectation value has components labeled ρ_s.[148] The mass matrices m_{sgl} and m_{Dsgl} are the complex constant Species mass matrix contributions for normal and Dark species.

[148] The Higgs fields $\eta\dots$ in our PseudoQuantum formulation are $\eta\dots = \varphi_{1\dots}(x) + \varphi_{2\dots}(x)$ as described earlier.

51.4 Species Group Higgs Masses Shows Inertial Mass Equals Gravitational Mass

In Blaha (2016h) we showed that a Complex General Relativity transformation can be factored into the product of a complex-valued transformation and a real-valued General Coordinate transformation. The set of complex valued transformations form a U(4) group that we called the General Coordinate Reality group. The analogous Internal Symmetry Species Group has gauge fields that undergo spontaneous symmetry breaking and generate contributions to all fermion masses.

Since fermion field masses are now sums of ElectroWeak Higgs contributions, Generation group Higgs contributions, Layer group Higgs contributions, and General Coordinate Species Group contributions, and since the Species Group Higgs fields appear in all fermion masses, the equality of inertial and gravitational mass is proven. The Species Group Higgs particles' equations set the mass scale of gravitational mass, and thereby of all Higgs mass contributions. The scale of inertial masses equals to the scale of gravitational masses. ***Since an expression cannot mix mass scales, the gravitational mass scale must be the same as the inertial mass scale.***

Inertial mass equals gravitational mass.

We have established the equality of inertial and gravitational mass at the short distance quantum level. In our view, this explanation is far more satisfying than basing the equality on a combination of large distance phenomena and quantum phenomena. As Einstein and Weyl have pointed out, all fundamental physics phenomena should be based on a local theory.

We have mapped Complex General Relativistic transformations consisting of U(4) transformations and Real General Relativistic transformations into transformations consisting of Internal Symmetry Species Group factors and Real General Relativistic transformations factors. The Higgs Mechanism breakdown of the Species Group has the important consequence that it prevents Species Group transformations that rotate between fermions and anti-fermions.

Appendix C. Some Features of QUeST Quaternion Space

This appendix describes 32 complex quaternion dimension space and some of the features of QUeST. It consists of material from Blaha (2020d) – (2020i).

C.1 Thirty-Two Complex Quaternion Space – 32 × 8 Dimension Array

A quaternion contains four dimensions. A complex quaternion contains eight dimensions. It is a complexification of the quaternion concept. Fig. C.1 depicts the 32 dimension space. It uses a "dot" • to represent a dimension. The dimensions of the space are not assigned physically until they are mapped to internal symmetry group fundamental representation dimensions and space-time dimensions. Rather than create a cumbersome coordinate-based notation we choose to use •'s.

•••• ••••
•••• ••••
•••• ••••
•••• ••••
•••• ••••
•••• ••••
•••• ••••
•••• ••••
...
•••• ••••

Figure C.1. The 32 complex quaternion dimension QUeST array. This array is the 32 × 8 array of •'s.

The 32 × 8 form of the array is useful because it brings out the four layers of fermions that appear when the array is subdivided into four layers (8 rows each) of fundamental group representations. The Unified SuperStandard Theory[149] implied by QUeST has a matching four layers of fermions. The subdivision of Fig. C.1 into layers appears in Fig. C.2. The map to group representations appears in Figs. C.3 and C.4. We use the maps in Table 1.1 to set up the group ↔ dimension map, bearing in mind the group representations of the Standard Model:

[149] Blaha (2020d) and earlier books.

U(1)	$\leftrightarrow$	2 real dimensions
U(4)	$\leftrightarrow$	8 real dimensions
U(2)	$\leftrightarrow$	4 real dimensions
SU(3)	$\leftrightarrow$	6 real dimensions
U(1)⊗SU(2)	$\leftrightarrow$	4 real dimensions

Table C.1. Map between fundamental representations and their dimensions.

where the dimensions have *real-valued* coordinates and are called *real dimensions*.

The eight U(4) Layer groups and eight U(4) Generation groups in Fig. C.3 are present in both UST and QUeST as well as being implied by BQUeST. See Fig. C.4 for the groups of one layer of the four layers.

C.1.1 QUeST Internal Symmetry Groups

The QUeST internal symmetry group is

$$[\mathrm{SU}(2)\otimes\mathrm{U}(1)\otimes\mathrm{SU}(3)]^{8}\otimes\mathrm{U}(4)^{16}\otimes\mathrm{U}(2)^{4} \tag{C.1}$$

by Fig. C.3 below.

C..1.2 QUeST Space-Time

The space-time of QUeST is 4 complex quaternion dimensions.

C.1.3 Generation and Layer groups of UST, QUeST and BQUeST

The Generation groups mix the fermion generations of normal and Dark sectors of each layer. The lines on the left side of Fig. C.5 display Generation group mixing within each layer.

Layer groups mix fermions in all four layers for each of the four generations individually. (See right side of Fig. C.5.) There are eight Layer groups: two Layer groups for Normal and Dark sectors for each generation.

The Dark groups mix between normal and Dark fermions, fermion by fermion.

C.1.4 Fermion- Dimension Duality

Fig. C.6 shows a 1:1 relation between QUeST dimensions and the fundamental fermions of QUeST and UST. This duality is the basis of the BQUeST and BMOST one-dimension theories implying QUeST and UTMOST respectively.

C.1.5 Fermion Structure Extracted from QUeST Symmetry Structure

Given the form of the internal symmetries in QUeST we can determine the fermions in the fundamental group representations as shown in Fig. C.5.

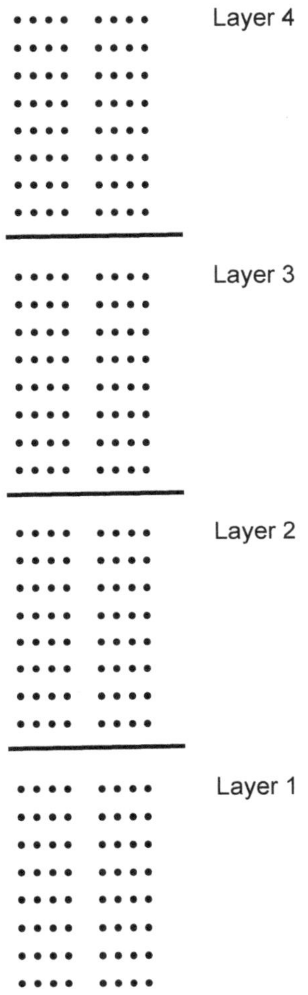

Figure C.2. The 32 complex quaternion dimension QUeST array subdivided into 4 layers of 8 rows..Each layer will be seen to map to a block of fundamental group representations as shown in Figs. C.3 and C.4.

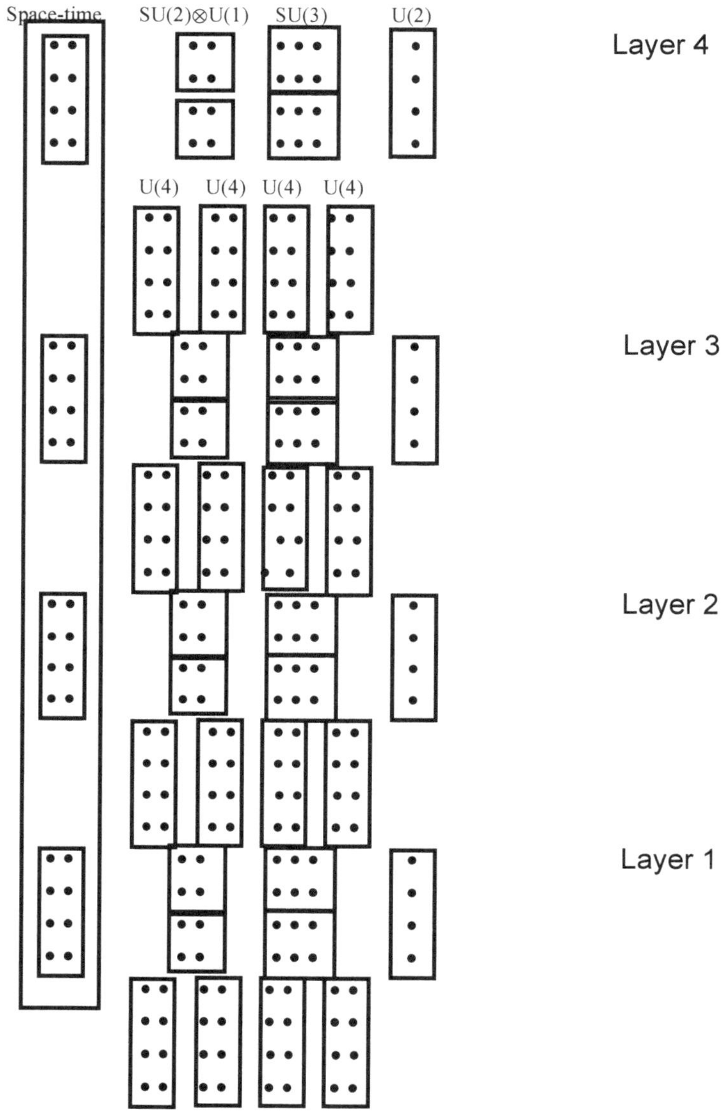

Figure C.3. The four layers of QUeST internal symmetry groups (and space-time) for 32 dimension complex quaternion space. Note: each row has an 8 • complex quaternion. Note the left column of blocks combine to specify a 4 dimension complex quaternion space-time. Note each layer requires 64 dimensions. *The U(2) Dark group is broken to U(1)⊗U(1) in chapter 2 to conform to the breakdown to 4 × 4 blocks suggested by comparison to the spinor structure.*

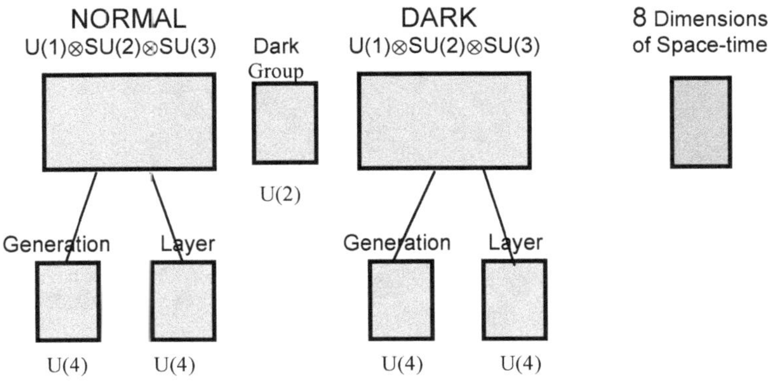

Figure C.4. The internal symmetry groups for one QUeST layer of the 4 layers in the 32 × 8 dimension array format. The two large blocks are each 5 complex dimension (10 real dimension) representations of SU(2)⊗U(1)⊗SU(3). The U(2) group (badly broken) supports transformations (rotations) between Normal and Dark matter.

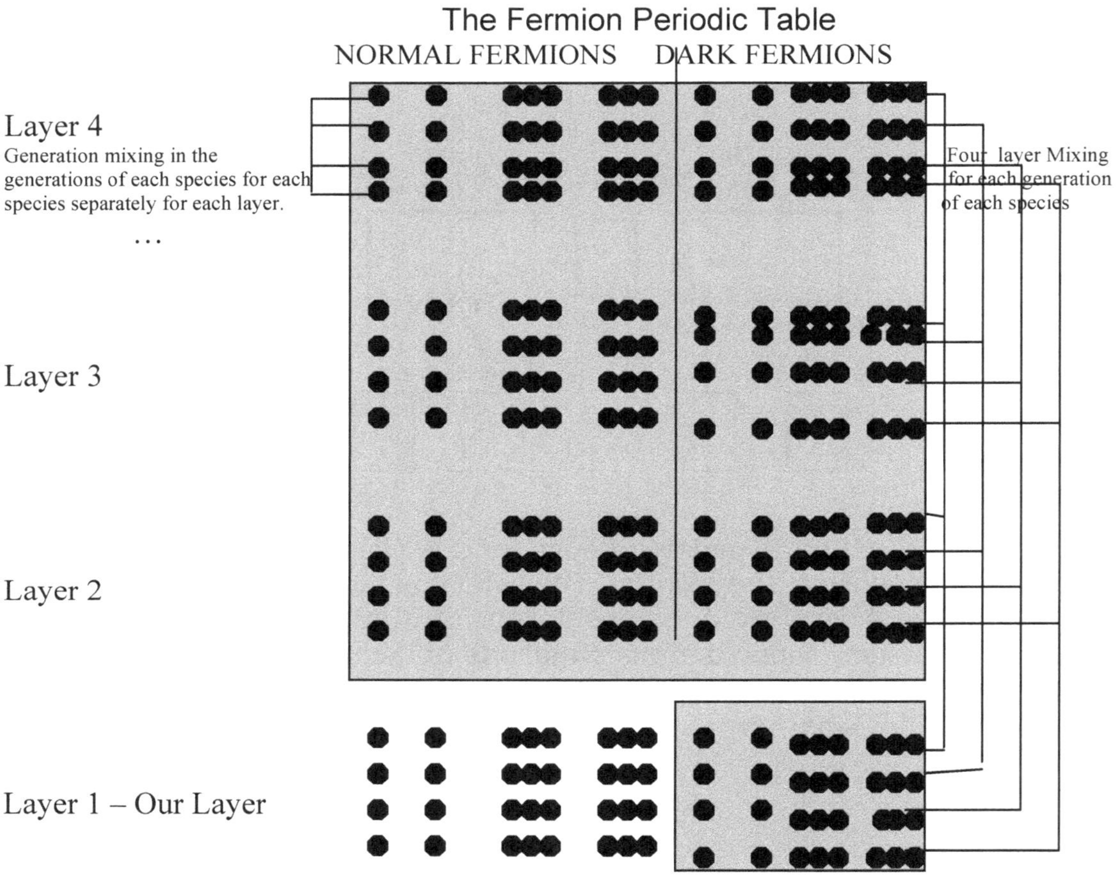

Figure C.5. Fermion particle spectrum and partial examples of the pattern of mass mixing of the Generation group and of the Layer group. Unshaded parts are the known fermions with an additional, as yet not found, 4th generation. The lines on the left side (only shown for one layer) display the Generation mixing within each layer. The Generation mixing occurs within each layer using a separate Generation group for each layer. The lines on the right side show Layer group mixing (for Dark matter) with the mixing among all four layers for each of the four generations individually. There are four Layer groups for Normal matter and four Layer groups for Dark matter.. There are 256 fundamental fermions. QUeST and UST have the same fermion spectrum.

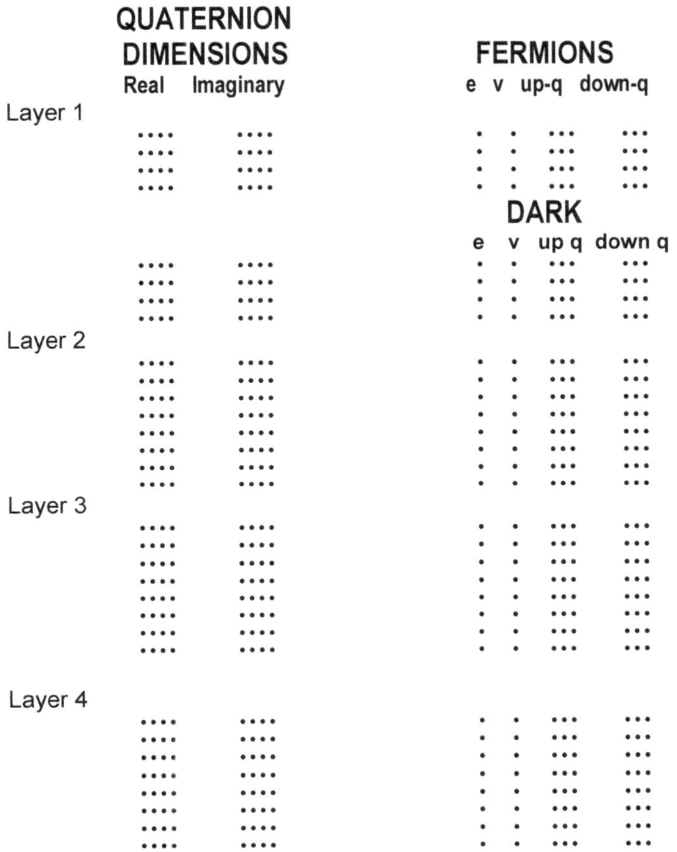

Figure C.6. Fundamental fermions have a 1:1 correspondence with QUeST dimensions. Note the number of dimensions in each row is 8 – the number of dimensions in a complex quaternion. Correspondingly the number of fermions in each row is 8 – a suggestive similarity. Each layer has four normal fermion generations and four Dark fermion generations. Each dot (pebble) represents a dimension in the left part of the figure and a fermion in the right part.

C.2 Thirty-Two Complex Quaternion Space – 16 × 16 Dimension Array

This section describes an alternate 16 × 16 form of the QUeST dimension array. This format supports the derivation of the QUeST dimensions from BQUeST.

The 16 × 16 form of the QUeST dimension array can be based on a 16 complex octonion dimension space. The difference between this format space and the 32 dimension space is not physically meaningful at present. The difference will be physically meaningful if the masses of the fermion spectrum and the full pattern of symmetry breaking is determined. Then one can differentiate between a four layer theory as above in section C.1 and a two "layer" theory presented below in this section.

The alternate 16 × 16 form of the QUeST dimension array is simply constructed from the preceding 32 × 8 dimension array by "moving" the 16 × 8 lower half of the 32 × 8 array of Fig. C.1 to the "right" of its upper half. The 16 × 16 dimension array appears in Fig. C.7. Note that the forms are physically equivalent if done prior to mapping dimensions to group representations.

The 16 × 16 form of the dimension array is more convenient for our derivation of a fundamental basis for QUeST that we call BQUeST (pronounced bee-quest). No change in the number of dimensions is made. (QUeST could also be viewed as residing in a 16 complex octonion dimension space for the purpose of dimension counting.)

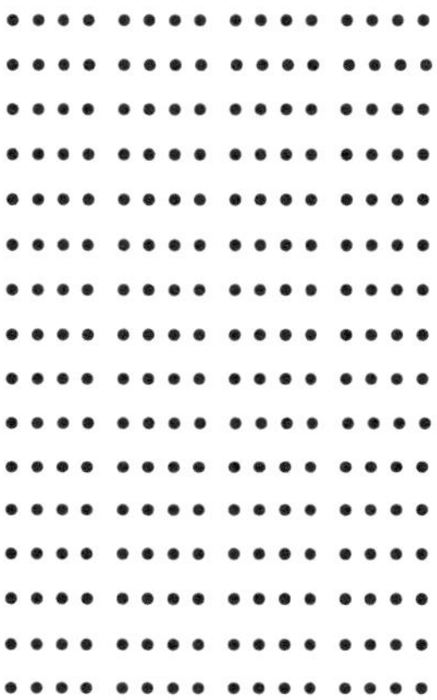

Figure C.7. The 16 × 16 array of QUeST dimensions.

Fig. C.8 shows the assignment of fundamental group representations to the 16 × 16 dimension array. Note this form of array has two "layers" of 128 dimensions in contrast to the four 64 dimension layers of Fig. C.3. The difference can be attributed to the third and fourth layers of Fig. C.3 becoming "extra" Dark sectors in "layers" 1 and 2.

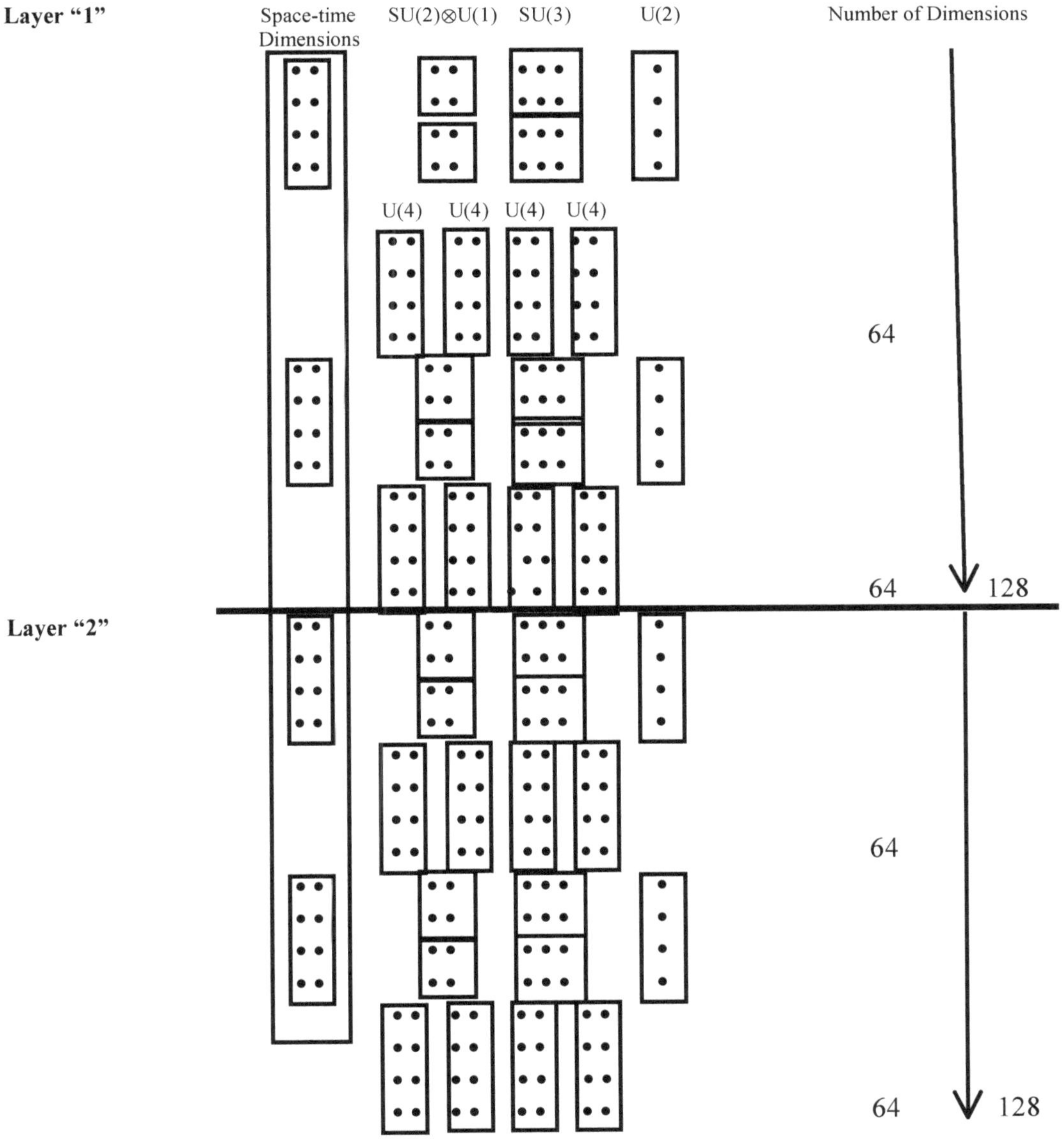

Figure C.8 The 16 × 16 form of QUeST array has two "layers", each of which is composed of two layers of the 32 × 16 four layer figure of Fig. C.3. Note each of these "layers" has 128 dimensions.

The group representations in Fig. C.8 can be subdivided into blocks of 16 dimensions as shown in Fig. C.9 and C.10 for the “layers” of Fig. C.8. Blaha (2020i) provides a rationale for 16 dimension subblocks based on U(8) considerations.

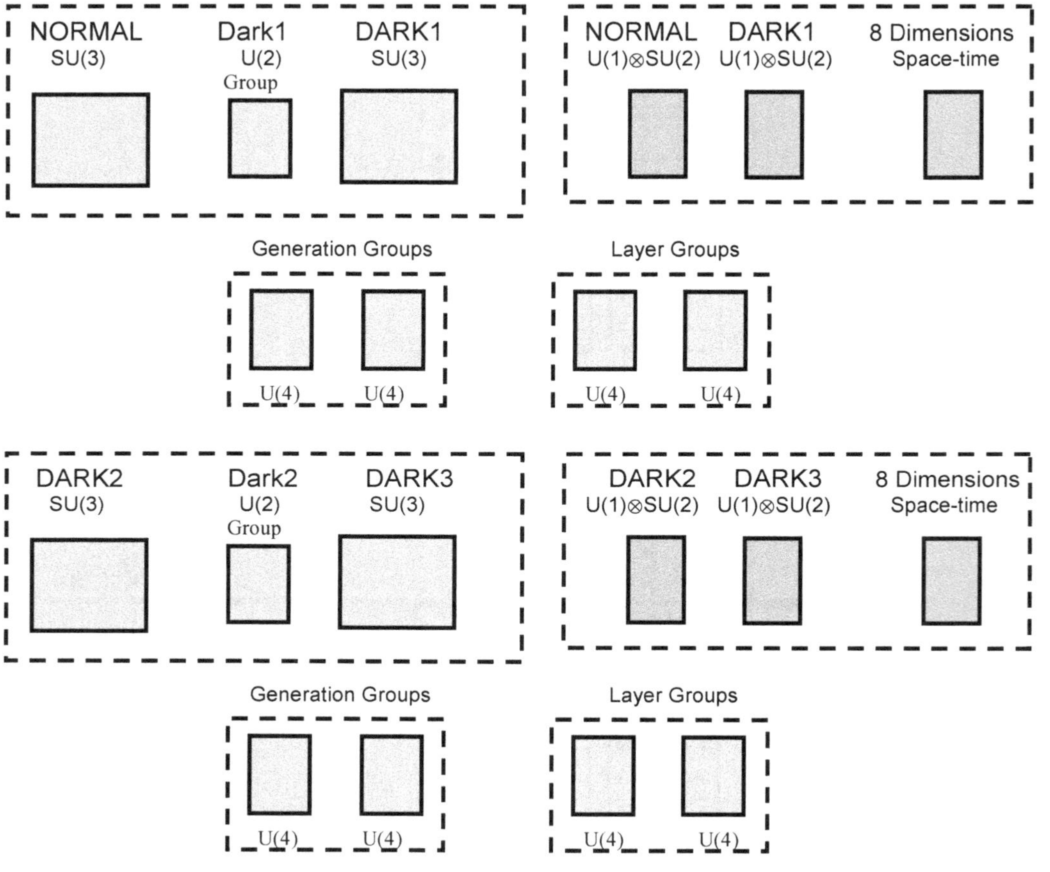

Figure C.9. Set of 16 dimension blocks for the first “layer” (of two “layers”) of the 16 × 16 array. Each “dashed” block (regardless of its apparent size) is a 4 × 4 = 16 array of dimensions. This set of 8 blocks contains the 8×16 = 128 dimensions of “layer” 1. “Layer” 2 is similar. The Dark U(2) groups supports transformations (rotations) between the types of matter: Normal and Dark1; and Dark2 and Dark3.

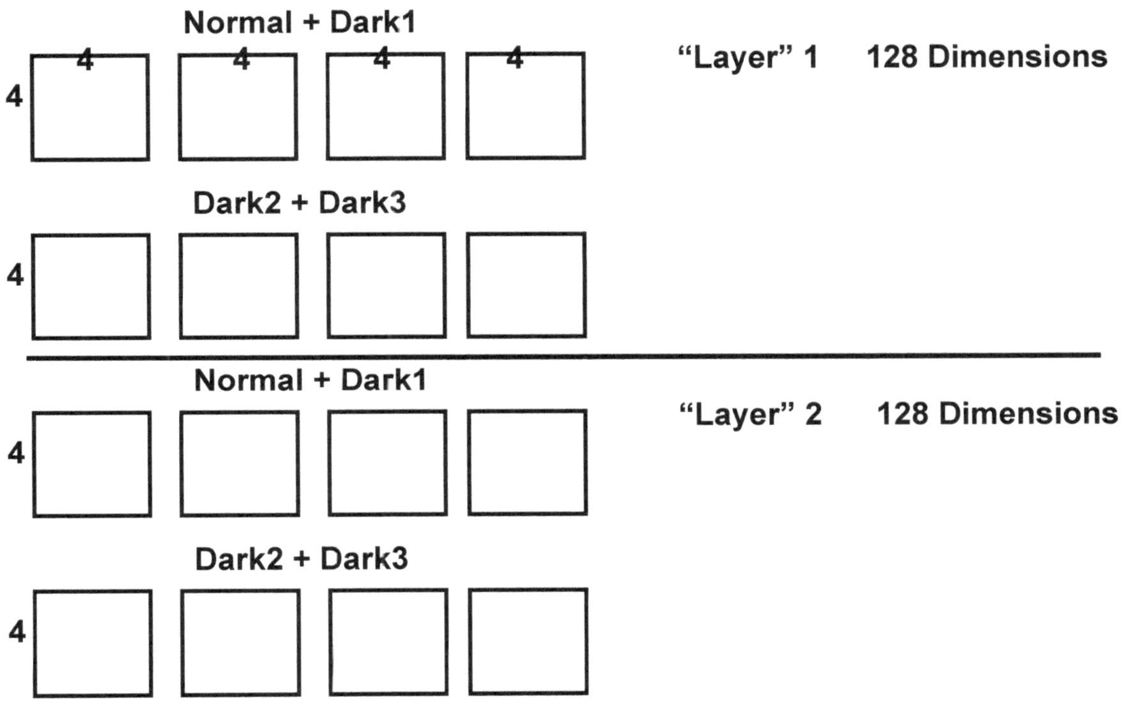

Figure C.10. The two "layers" of 4 × 4 dimension subblocks of the 16 × 16 dimension array. .

QUeST fermions have a similar format to the dimension subblock structure in Fig. C.10. Fig. C.11 displays the 256 fermions of QUeST. The shift from a 32 × 8 dimension array with four layers of groups to a 16 × 16 dimension array gives a two "layer" form with the" lower" two layers of the 32 × 8 dimension array becoming the Dark2 and Dark3 parts of the two "layer" form. This shift has no discernable physical consequences as far as our considerations are concerned since we are moving Dark sectors. When mass generation and symmetry breaking are better understood, a physically significant difference may be evident.

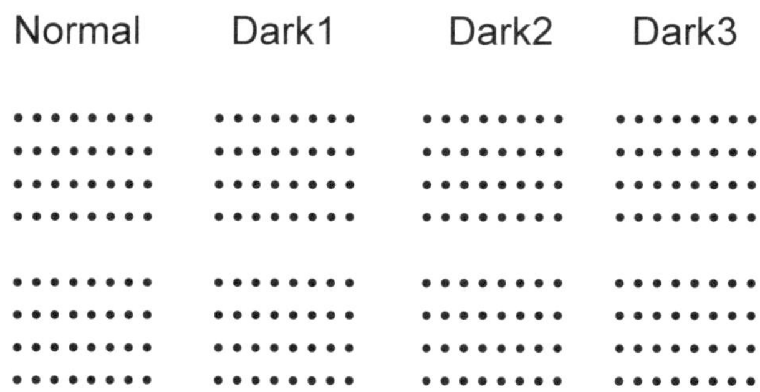

Figure C.11. Spectrum of the generations of fermions of QUeST for the 16 × 16 dimension array representation. Each fermion is represented by a •. Quark triplets are represented by three •'s. Note there are 256 fundamental fermions.

The possible U(8) 4 × 4 block structure of Figs. C.9 – C.10 suggest a similar block structure for the fundamental fermions. Fig 1-A.12 displays a set of sixteen 4 × 4 blocks with each block holding 16 fermions. It has a SU(4)-like structure of the four rows (generations) of each 16 dimension subblock. The subblocks have a lepton-triquark content. Clearly the implied SU(4) symmetry would be broken. It is suggestive of a Lorentz 4-vector representation with the lepton corresponding to a time coordinate and the three quarks corresponding to spatial coordinates.

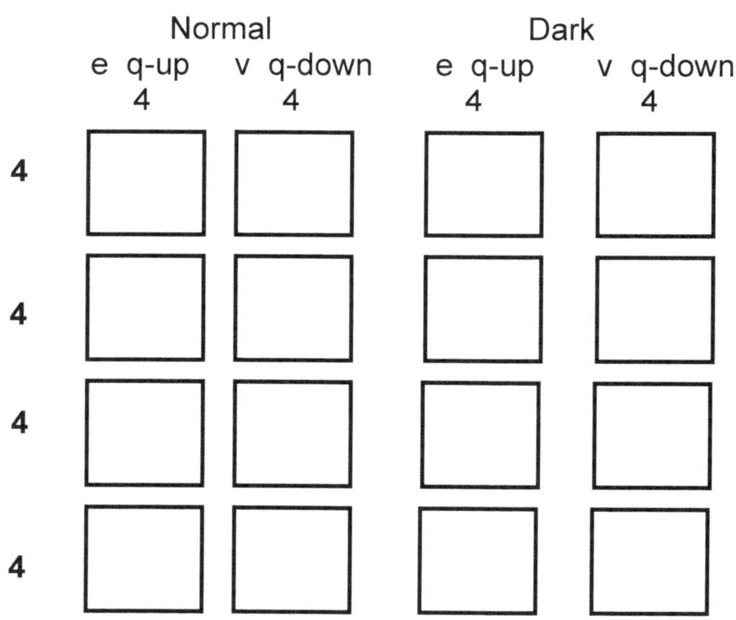

Figure C.12. Block form of a 16 × 16 QUeST fermion array with each block row corresponding to one layer. Each block contains four generations of fermions. The result is 4 × 4 blocks. The label e q-up indicates a charged lepton – up-type quark pair, ν q-down indicates a neutral lepton – down-type quark pair, and so on. Note the blocks can be reaaranged into a 32 × 8 form without physical consequences at this level of discussion since the right two columns and the lowest two rows are all Dark at present.

C.3 A Partition to Real 3+1 Dimension Space-times

We can partition Fig. C.4 into real and imaginary subspaces with 3+1 dimension space-times.

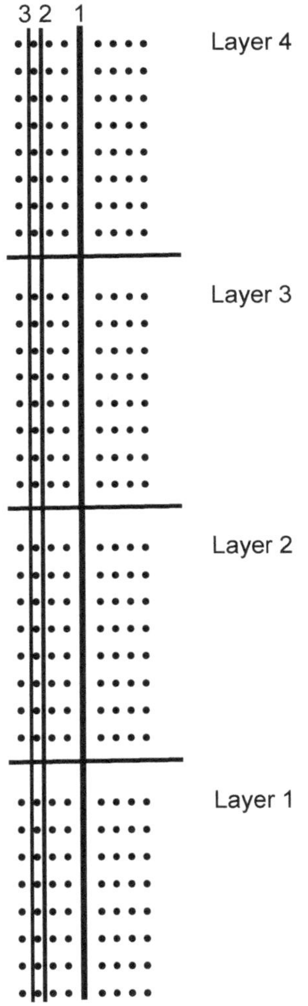

Figure C.13. The 32 complex quaternion dimension QUeST array partitioned down to a real spwce-time. The partition labeled "1" reduces the array to 32 quaternion dimensions after discarding the right columns. The partition labeled "2" reduces the array to 32 complex dimensions similarly. The partition labeled "3" reduces the array to 32 real-valued dimensions in the leftmost column..

The real-valued column of dimensions numbers 32 dimensions. It maps to the internal symmetry group:

$$[SU(2)\otimes U(1)\otimes SU(3)]^2\otimes U(4) \qquad (C.2)$$

where the U(4) group is the Generation group, which implies 4 generations. There is an SU(2)⊗U(1)⊗SU(3) group giving the ElectroWeak and Strong interactions of Normal matter. The other SU(2)⊗U(1)⊗SU(3) group gives the Dark Matter ElectroWeak and Strong Interactions. Note the Layer group is not present.

These groups require 28 dimensions. The remaining four dimensions give 4 dimension space-time. The total is 32 dimensions.

The fundamental fermions corresponding to the internal symmetry groups number 32 fermions (fermion-dimension duality). See Fig. C.14.

Figure C.14. Spectrum of a generation of fermions. Note this is only one layer of fermions. Two leptons and 6 quarks in each row of Normal matter. Similarly for Dark matter.

This restricted model is similar to The Standard Model: same internal symmetry, same 4 generations, same space-time, and same fermion and vector boson spectrums..

Appendix D. Evidence for Universe Particles

This appendix provides evidence for the existence of a Megaverse and for the possibility that universes are particles. It is abstracted from Blaha (2018e), which provides much more detailed data.

D.1 Evidence for Entities Beyond the Universe

At first glance it would seem impossible to produce evidence for the existence of other universes. However there are subtle means by which we can 'sense' experimentally 'nearby' universes should they exist. The mechanism would appear to be gravitational effects exerted on objects within our universe by unseen objects of enormous mass. Currently there appears to be three experimental suggestions of the existence of 'nearby' universes and one theoretical argument based on an influx of mass-energy from the Megaverse that may cause the expansion of our universe.

There are also theoretical motivations for believing that there are entities beyond our universe. These are detailed in chapter 30 of Blaha (2018e).

D.1.1 Great Attractors

One potential support is the discovery of the Great Attractor (at the center of the Laniakea Galaxy Supercluster), and the more massive Shapley Attractor (centered in the Shapley Supercluster)[150]. These attractors contain massive numbers of galaxies and are drawing galaxies over a distance of millions of light years towards them.

If another universe(s) is 'near' our universe it could act as a 'gravitational magnet' and draw galaxies within our universe towards it to form one or more superclusters which could then act as attractors. Thus attractors might indirectly reveal the presence of other nearby universes—contrary to the expected large scale uniformity of the universe. The only other apparent source of superclusters is chance. Chance seems an unsatisfactory possibility in the present case.

D.1.2 Bright Bumps in Universe Suggesting Collision with Another Universe

A recent study[151] of the residual brightness of parts of the accessible universe found that bright patches appeared if a model of the CMB (Cosmic Microwave Background) with gases, stars and dust was 'subtracted' from the PLANCK map of the entire sky. After the subtraction one would expect only noise spread throughout the sky. However, bright patches were seen in a certain range of frequencies. These anomalies are thought to be a result of our universe colliding with another object – presumably another universe in the Megaverse.

---

150 Tully, R. Brent; Courtois, Helene; Hoffman, Yehuda; Pomarède, Daniel, "The Laniakea Supercluster of galaxies". Nature (4 September 2014). 513 (7516): 71–73; arXiv:1409.0880.

151 Ranga-Ram Chary, arXiv.org:/1510.00126 (2015).

D.1.3 Cold Spot in Universe Suggesting Collision with Another Universe

Another recent study[152] of a huge cold region of the universe spanning billions of light years revealed that this region is not a relatively empty region but rather is similar to in its distribution of galaxies to the rest of the universe. Previous the Cold Spot (an area where cosmic microwave background radiation – the leftover Big Bang radiation is weak – making it significantly colder (0.00015C colder) than the average temperature of the universe.)

An analysis of 7,000 galaxy redshifts using new high-resolution data has now shown that the Cold Spot is similar to the rest of the universe. The Durham University group suggested that the Cold Spot might have been caused by a collision between our universe and another Universe. They further suggested that there is only a 1 in 50 chance that it could be explained by standard cosmology.

Thus we have another important piece of circumstantial evidence in favor of other universes and thus the Megaverse.

D.1.4 Conclusion

We conclude that data is beginning to emerge favoring multiple universes and a physical Megaverse in support of the theoretical justifications presented earlier.[153]

---

[152] T. Shanks et al, Durham University (Australia), Monthly Notices of the Royal Astronomical Society, 2016 .
[153] Chapter 59 of Blaha (2020c).

Appendix E. Some Features of UTMOST

This appendix describes the 64 complex octonion dimension space of UTMOST. It also discusses an alternate 32 *quaternion octonion*[154] dimension space for UTMOST. Given the present state of physical data these variants are both acceptable. and features of QUeST.

E.1 UTMOST Space

An octonion contains eight dimensions. A complex octonion contains sixteen dimensions. A *quaternion octonion*[155] contains 32 dimensions. Fig. E.1 depicts the 64 dimension complex octonion space as a 64 × 16 array of dimensions. It uses a "dot" or pebble • to represent a dimension[156]. The dimensions of the space are not assigned physically until they are mapped to internal symmetry group fundamental representation dimensions and space-time dimensions. Rather than create a cumbersome coordinate-based notation we choose to use •'s.

•••••••• ••••••••
•••••••• ••••••••
•••••••• ••••••••
•••••••• ••••••••
•••••••• ••••••••
…
•••••••• ••••••••

Figure E.1. The 64 complex octonion dimension UTMOST array. This is the 64 × 16 array of •'s. It has 1024 dimensions.

The UTMOST space can also be viewed as a space of 32 quaternion octonion dimensions. It also has a total of 1024 dimensions. Fig. E.2 creates a 32 × 32 array of dimensions for this space. *A 32 × 32 array is important for the derivation of UTMOST from one-dimension BMOST.*

The 32 × 32 form of the UTMOST dimension array is based on a 32 quaternion octonion dimension space. The difference between this form of space and the 64 complex octonion dimension space above is not physically meaningful at present. The

---

[154] A quaternion octonion is a 32 dimension set of coordinates. It is the composition of a quaternion and an octonion that could be represented by an expression such as q(o), which represents an octonion argument o for a quaternion functional q that expands each dimension in the octonion fourfold It is a generalization of a complex octonion..

[155] This notation, originated here, is for the composition of a quaternion and an octonion.

[156] The use of pebbles is a useful simplification of coordinates that enables assignments of pebbles (coordinates) to group representations to be visual.

difference will be physically meaningful if the masses of the fermion spectrum and the full pattern of symmetry breaking are determined. Then one can differentiate between the symmetry group spectrum and mass spectrums of the respective possibilities.

Figure E.2. The UTMOST array with 32 × 32 dimensions for a 32 quaternion octonion dimension space.

The repetitive pattern of groups seen in QUeST leads us to assume that UTMOST has a similar repetitive pattern. We will use a four layer format for the 32 × 32 array of dimensions. Each layer consists of 8 rows of Fig. E.2. Each layer can be put in a form analogous to Fig. C.4 (and to Fig. C.9). See Fig. E.3.

We map between dimensions and fundamental group representations. We use the maps in Table E.1 to set up the group ↔ dimension map, bearing in mind the group representations of the Standard Model:

U(4)	↔	8 real dimensions
U(2)	↔	4 real dimensions
SU(3)	↔	6 real dimensions
U(1)⊗SU(2)	↔	4 real dimensions

Table E.1. Map between fundamental representations and their dimensions.

Fig. E.3 shows the content of one UTMOST layer. The four layers of UTMOST are four copies of Fig. E.3.[157] The separation of the set of dimensions is accomplished by following the procedure given earlier.[158]

Fig. E.5 shows the four layers (each in two rows) of the 32 × 32 dimension UTMOST array, which is composed of 4 × 4 blocks. The 4 × 4 blocks are within the

[157] The Layer groups are U(4) groups. They mix the generations of each of the top four layers, generation by generation, separately from the Layer groups mixing the lower four layers. This feature enables QUeST universes to be generated from either the top four layers or the lower four layers.

[158] The separation of the dimensions into the subgroup factors' representation can be implemented as group transformations and definitions using standard group theoretic methods. A more formal method for extracting the subgroup content of representations uses a symmetric group analysis of U(n) representation characters. See S. Blaha, J. Math. Phys. **10**, 2156 (1969) for a detailed discussion of this approach.

four block 8 × 8 sections for each pair: Normal+Dark1, Dark2+Dark3, Dark4+Dark5 and Dark6+Dark7. In total they form the 32 × 32 = 1024 UTMOST dimension array.

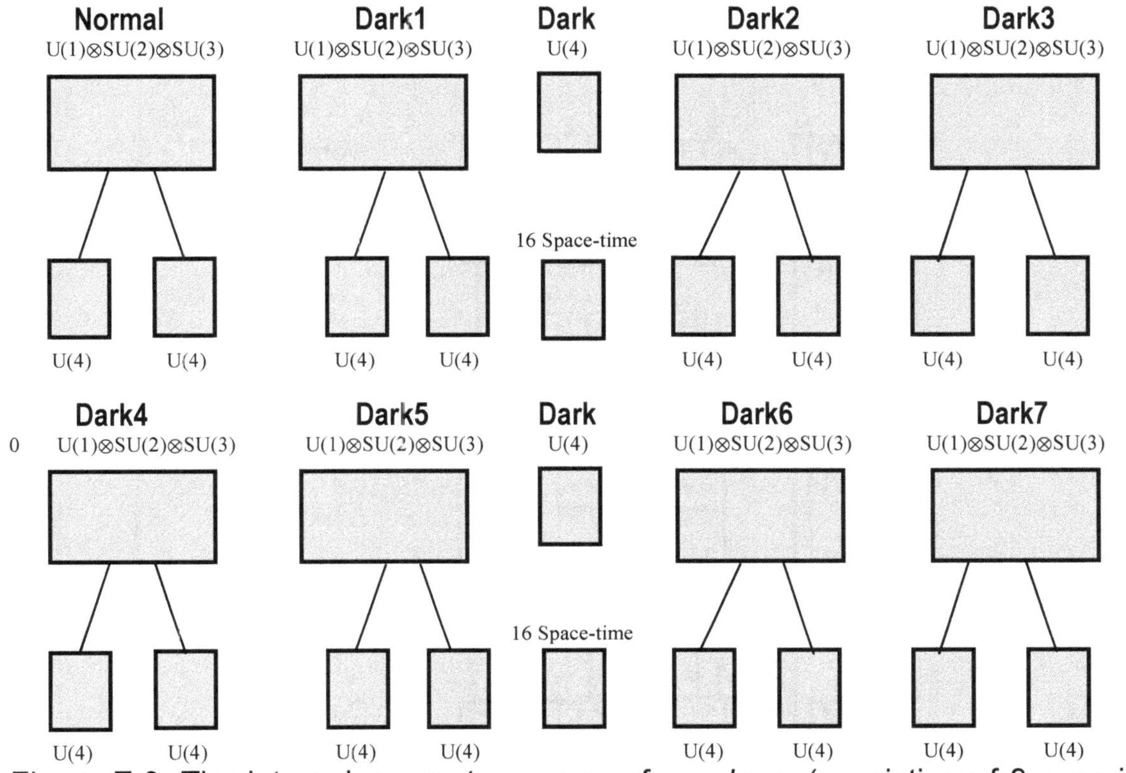

Figure E.3. The internal symmetry groups of *one layer* (consisting of 8 rows in Fig. E.2) of the four layers of 32 × 32 dimension UTMOST. The other three layers are copies of the this layer. Note the Dark U(4) groups. One U(4) "rotates" among Normal, Dark1, Dark2, and Dark3. The other U(4) "rotates" among Dark4, Dark5, Dark6, and Dark7. The U(4) group may be broken to U(1)⊗U(1)⊗U(1)⊗U(1) to obtain 4 × 4 blocks.

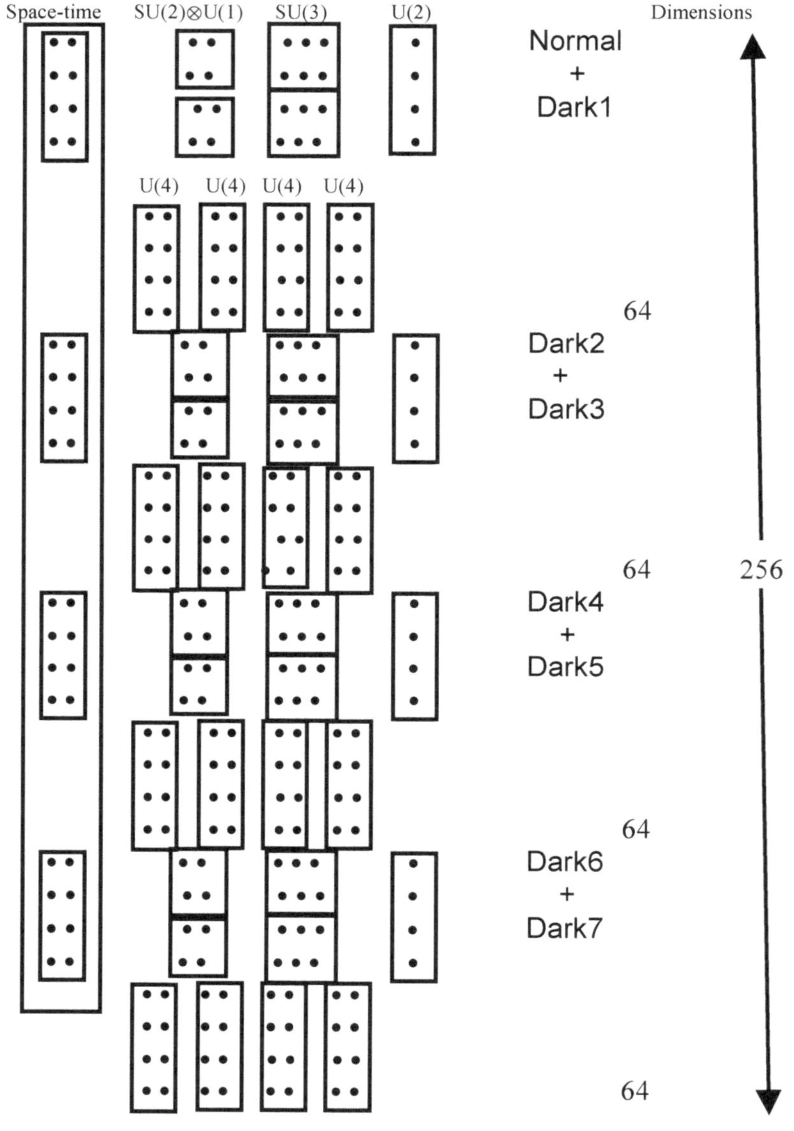

Figure E.4 The *first* of the four layers of UTMOST dimensions with boxes around sets of dimensions for fundamental group representations. The U(4) Dark groups have been broken into U(2)⊗U(2) factors. The U(4) group may be further broken to U(1)⊗U(1)⊗U(1)⊗U(1) to obtain 4 × 4 blocks.

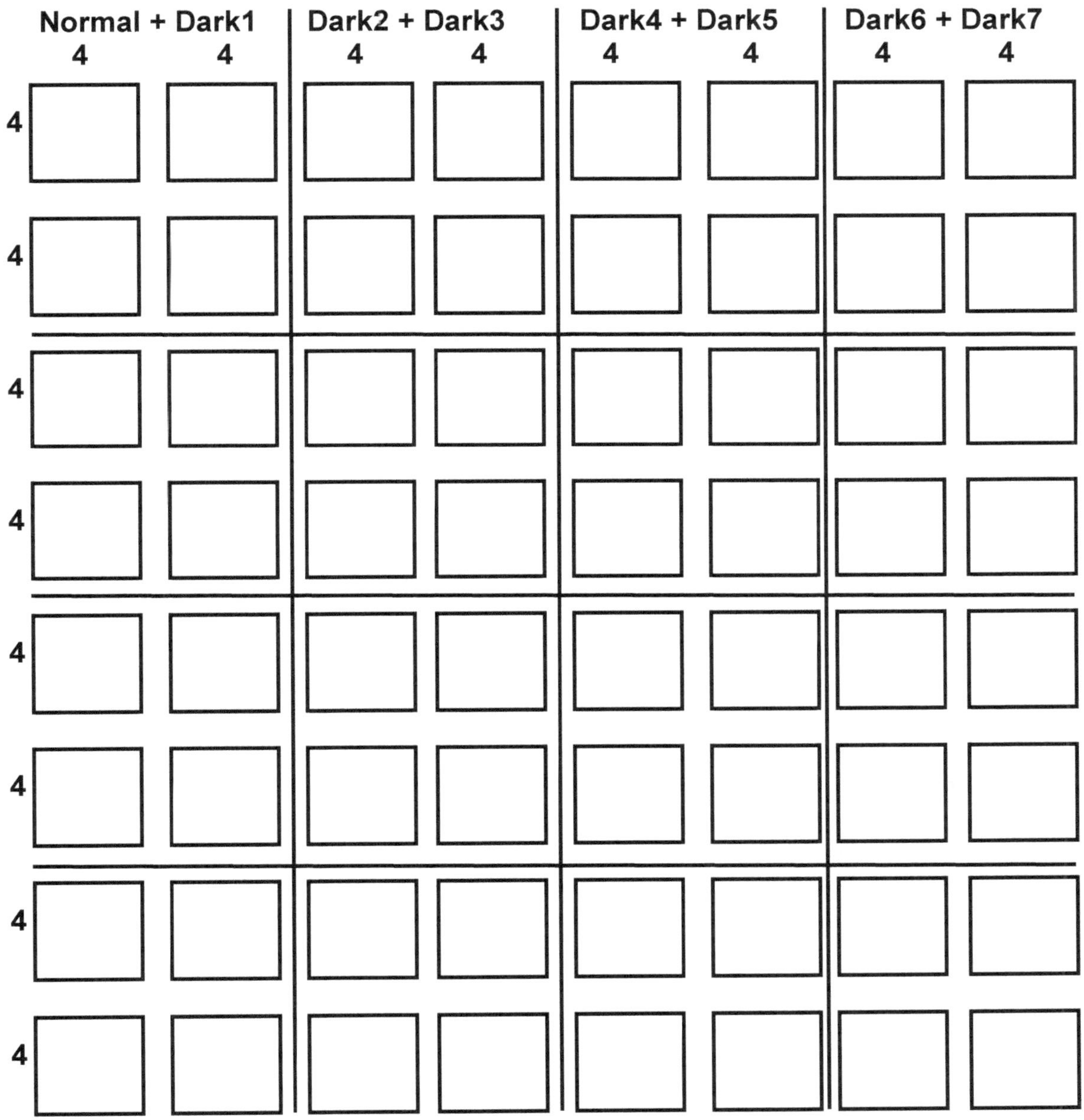

Figure E.5. Four layers (each in two rows) in the 32 × 32 dimension UTMOST array composed of 4 × 4 blocks, which are within the four block 8 × 8 sections for each pair: Normal+Dark1, Dark2+Dark3, Dark4+Dark5 and Dark6+Dark7. In total they form the 32 × 32 = 1024 UTMOST dimension array.

E.2 UTMOST Fermions

Given the form of the internal symmetries in UTMOST we can determine the fermions in the fundamental group representations as shown in Fig. E.6.

UTMOST Fermion Array

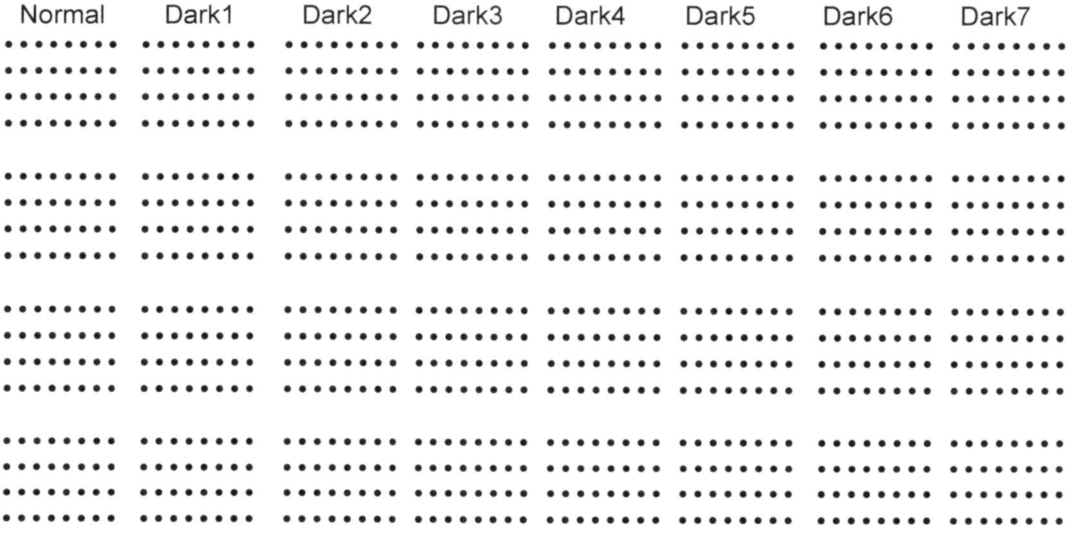

Figure E.6. Spectrum of UTMOST fermions in a 16×64 format. Each fermion is represented by a •..Each set of eight •.'s represents a charged lepton, a neutral lepton, three up-type quarks, and three down-type quarks. There are eight sets of four species in four generations which are in turn in 4 layers. There are 1024 fundamental fermions taking account of quark triplets.

In chapter 3 we outlined possible patterns of subspaces of QUeST and UTMOST. One choice of pattern is based on 4 × 4 blocks of dimensions, assembled into 8 × 8 blocks of dimensions containing four 4 × 4 blocks, assembled in four layers. Fig. E.7 shows the possible implications of this arrangement for fermions. The 4 × 4 fermion blocks contain either four generations of charged leptons and up-quarks, or four generations of neutral leptons and down-quarks.

The grouping of a lepton and three quarks in both cases creates a similarity to time and spatial coordinates respectively suggesting a broken Lorentz group-like structure or a possible SU(4) broken symmetry.

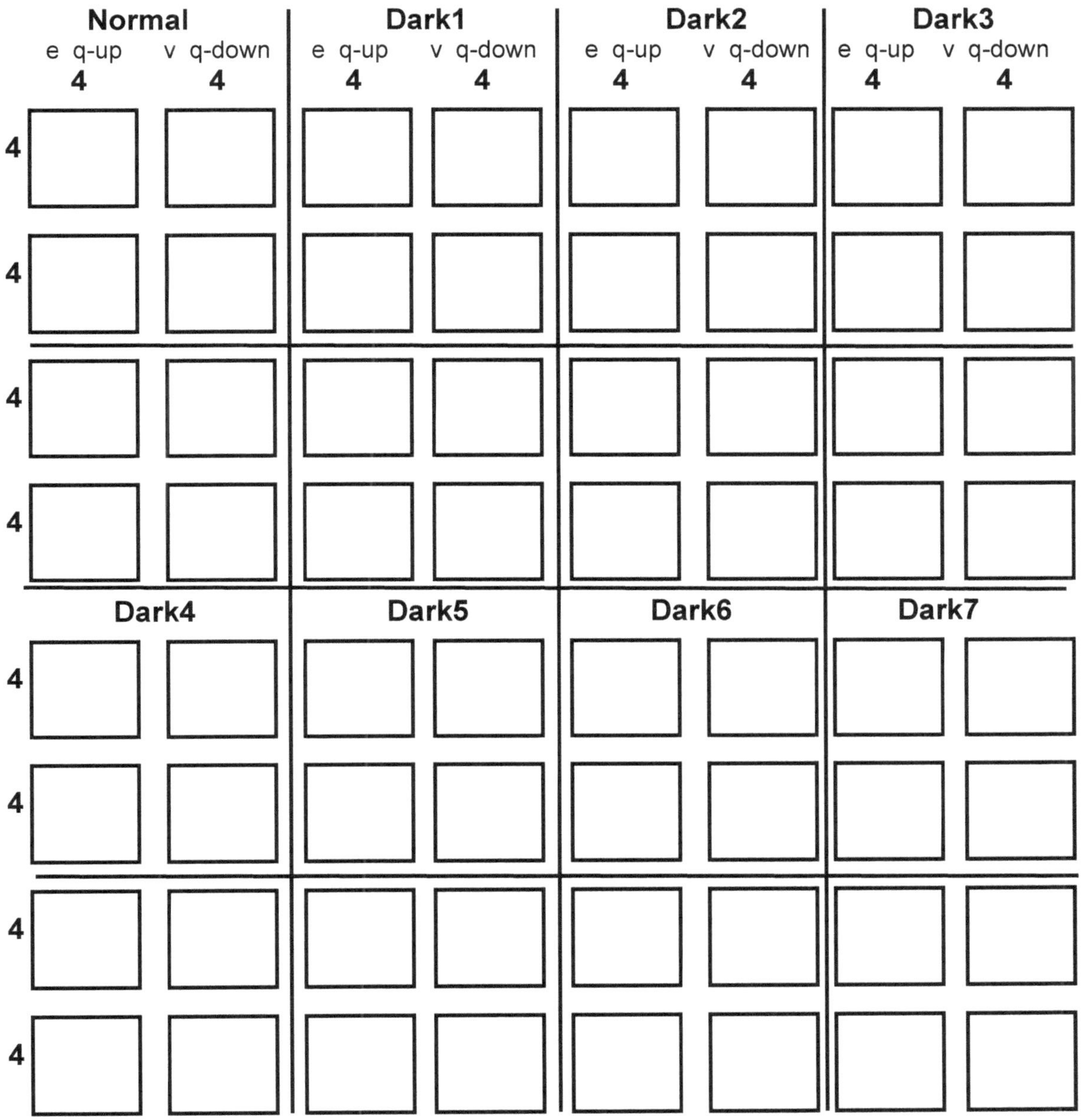

Figure E.7. Block form of the 32 × 32 UTMOST fermion array with each row corresponding to *half of an UTMOST layer*. Thus 8 × ½ = 4 layers results. Each block contains four generations of fermions. The result is sixty-four 4 × 4 blocks. The label e q-up indicates a charged lepton – up-type quark pair, v q-down indicates a neutral lepton – down-type quark pair, and so on. The form displayed here may explain why generations come in fours.

E.3 Partition of UTMOST into QUeST Subspaces

UTMOST[159] can be partitioned into several levels of subspaces. We modify the UTMOST figures seen earlier to show a partition into two subspaces. These spaces will be MOST spaces.[160] Then we can partition a MOST subspace into two QUeST subspaces and so on.

The partitioned UTMOST dimension arrays are:

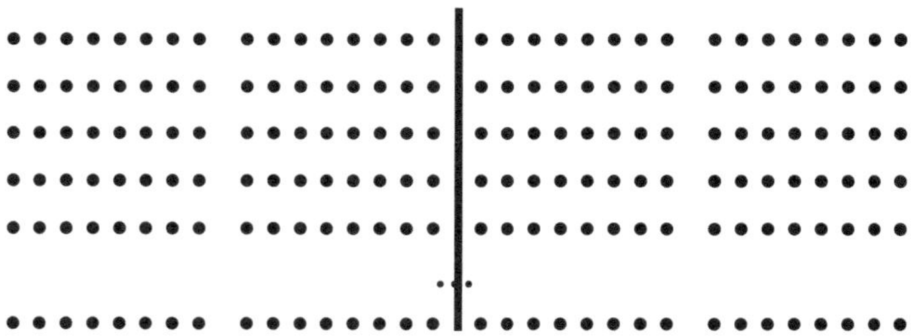

Figure E.8. The partition of the 32 × 32 dimension UTMOST array into MOST subspaces The size of each subspace is 32×16 = 512 dimensions.

The partition of UTMOST fermions into MOST fermions appears in Fig. E.9.

[159] We will consider MOST subspaces below.
[160] MOST is described in Blaha (2020i) and earlier books by the author.

UTMOST Fermion Array Partitioned into MOST Fermion Arrays

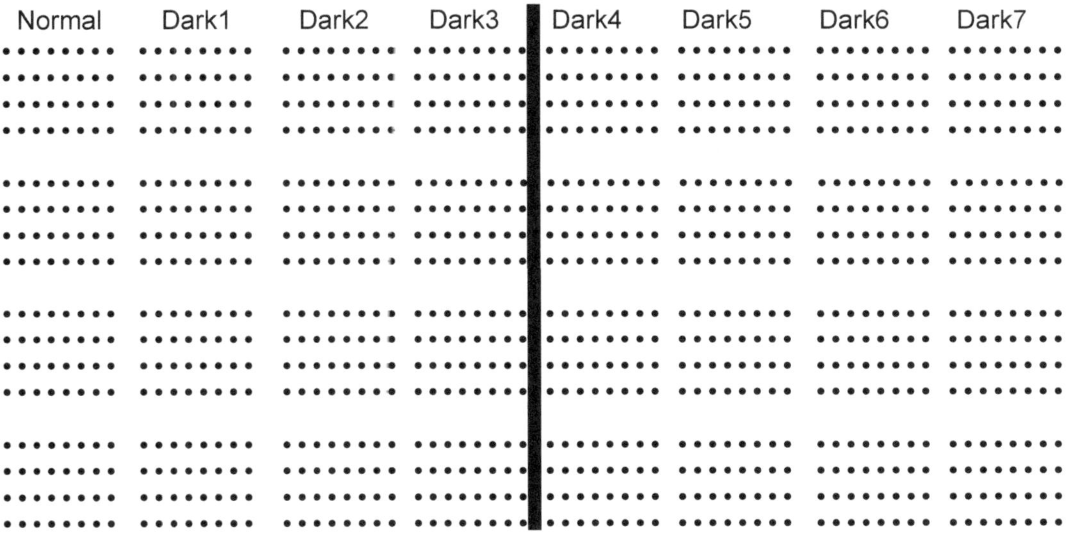

Figure E.9. Partition of spectrum of UTMOST fermions in a 16×64 format. Each fermion is represented by a •. Including each quark. Each set of eight •.'s represents a charged lepton, a neutral lepton, three up-type quarks, and three down-type quarks. There are eight sets of four species in four generations which are in turn in 4 layers. There are 512 fundamental fermions in each subspace taking account of quark triplets. Note: Quark singlets won't do; triplets are required.

UTMOST Fermion Array Partitioned into MOST Fermion Arrays with 4 × 4 Blocks

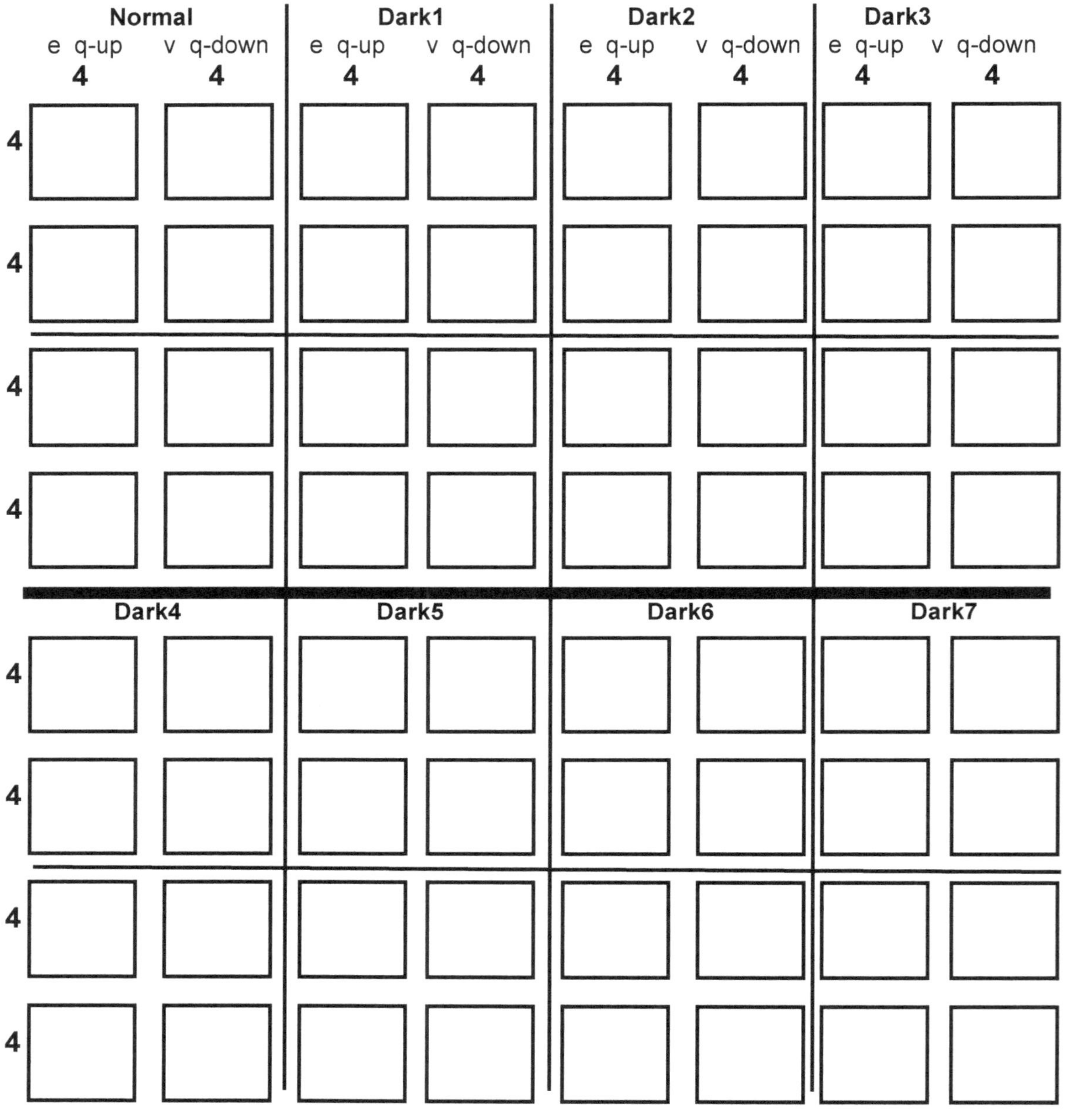

Figure E.10. Partition of block form of the 32 × 32 UTMOST fermion array with each row corresponding to *half of an UTMOST layer*. Thus 8 × ½ = 4 layers results. Each block contains four generations of fermions. The result is sixty-four 4 × 4 blocks. The label e q-up indicates a charged lepton – up-type quark pair, v q-down indicates a neutral lepton – down-type quark pair, and so on.

Appendix F. "D" Dimension Quantum Field Theory from Chapter 10 of Blaha (2017c)

This appendix describes D dimension Quantum Field Theory. It appeared in chapter 10 of Blaha (2017c). The case of the BQUeST seed fermion requires dimension D = 8. The seed urfermion for BMOST requires D = 10. A 16 dimension quantum field theory version appears in Blaha (2014a).

10.2 The Particle Interpretation of Extended Wheeler-DeWitt Equation Solutions

In earlier chapters we described features of the Wheeler-DeWitt equation that suggested that universes could be viewed as particles or anti-particles, or tachyons. The solutions of this equation are scalar wave functions on a manifold that are analogous to the solutions of the Klein-Gordon equation. The issues of negative probabilities, possible tachyonic solutions, and negative frequency solutions suggest a need for an appropriate particle interpretation of universes that can possibly resolve these problems.

Some physicists have taken the Wheeler-DeWitt equation as the starting point for a theory of a universe as a particle. The Wheeler-DeWitt equation describes the interior of a universe in a quantum framework.

We will take a different approach using the Megaverse as the environment of universe particles that internally have Quantum Gravity, and externally have Megaverse Quantum Gravity.

We view a universe as an extended particle and begin by ignoring the detailed inner structure of universes. This approach is similar to the historical treatment of hadrons such as the proton as particles and developing a theory of them as fundamental particles using form factors, structure functions and so on to approximate their inner structure. Afterwards, as detailed data became available, the detailed investigation of the internal structure of hadrons using quark-parton models followed. We will pursue a similar theoretical development beginning with a theory of universes as extended particles in the D-dimension Megaverse. The internal structure of the particle universes will eventually be specified by the Wheeler-DeWitt equation expressed in Megaverse coordinates.

The two simplest choices for the nature of universes are "spin 0" *bosonic universes* and *fermionic universes* with odd half integer spin, s_M.[161] We will first consider the possibility of fermionic universes, and then briefly consider "spin 0" *bosonic* universes.

The first issue of fermionic universes (reminiscent of the discussions of spin in the 1920's) is the interpretation of spin states. We suggest that the upper $2^{D/2-1}$

[161] Since the Megaverse is D-dimensional, the spin of fermionic universe particles was shown to be s_M.

components (with $2^{D/2-2}$ "spin up" and $2^{D/2-2}$ "spin down" states) of a fermionic universe wave function represent a left-handed universe with an excess number of baryons. The lower ($2^{D/2-1}$) components lead to right-handed anti-universes where there is an excess of anti-baryons. These associations are analogous to the interpretations of the Dirac electron wave function.[162]

The universe particle "spin up" and "spin down" states are distinguished by their interactions with gauge fields in a manner analogous to quantum electrodynamics.

10.3 "Free Field" Dynamics of Fermionic Universe Particles

We now consider universes as extended particles with an odd half integer spin – *fermionic universe particles* - in the D-dimensional Megaverse. In the Megaverse there are D 'Dirac' matrices with $2^{D/2}$ rows and $2^{D/2}$ columns that are the equivalent of the four Dirac matrices in four dimensions. We will denote these D matrices as $\gamma_M{}^i$ for i = 1, 2, … , D. They satisfy the anti-commutation relations:

$$\{\gamma_M{}^i, \gamma_M{}^j\} = 2\,\delta^{ij} \tag{10.1}$$

and thus form a Clifford algebra. We will choose y^D to be the time coordinate and thus make it pure imaginary with a Reality group transformation. (The D-dimensional Megaverse space is a complex Euclidean space.) Therefore γ^D will be hermitean ($(\gamma^D)^2 = 1$), and the γ^i matrices for i = 1, … , (D – 1) will be anti-hermitean with $(\gamma^i)^2 = -1$. The number of linearly independent matrices in D dimensions is 2^D.

The Megaverse metric is (by use of the Reality group) chosen to be

$$g^{ij} = -\delta^{ij}, \qquad g^{D,D} = 1 \tag{10.2}$$

for i, j = 1, 2, … , (D – 1); and zero otherwise.

Except for the additional dimensions, fermion dynamics is quite similar to the 4-dimensional case. The free universe particle Dirac equation is

$$(i\gamma^i\partial_i + m)\psi(y) = 0 \tag{10.3}$$

summed over i = 1, 2, … , D where the mass is assumed to be constant, and set by eq. 10.119 below. The derivative operator, is based on the use of quantum coordinates[163] (eq. 7.9)

$$Y^i(y) = y^i + i\,Y_u{}^i(y)/M_u{}^{D/2} \tag{10.4}$$

For i = 1, …, D and is defined to be

[162] It is known that phenomena in our universe tend to be left-handed. If this feature of our universe's phenomena is also a property of the universe itself, then, since handedness is an attribute of spin, the treatment of a universe as having spin is not unreasonable.

[163] Giving Two Tier renormalization.

$$\partial_i = \partial/\partial Y^i(y) = \partial/\partial(y_i - Y_{ui}(y)/M_u^{D/2}) \qquad (10.5)$$

where *we assume $M_u = M_c$ with M_c being a very large mass scale of perhaps the order of the Planck mass.*

$Y_u^{\,i}$ is a D-dimensional Megaverse gauge field equivalent of the universe $Y^\mu(x)$ used in Two Tier renormalization (discussed in I):

$$Y^\mu(z) = z^\mu + i\, Y^\mu(z)/M_c^2$$

where $Y^\mu(z)$ is a free QED-like field. The $Y^i(y)$ quantum coordinates will be used in the Megaverse to eliminate potential divergences, in a manner similar to the case of our universe when universe particle interactions are introduced later.

10.3.1 Four Types of Fermionic Universe Particles

Assuming universe energies are real-valued,[164] there are four possible types of fermionic universe particles in the Megaverse that are analogous to the four species of fermion described in I (and Blaha (2010b)) for The Extended Standard Model. Two of these types are tachyonic. It is important to note that DeWitt points out that the Wheeler-DeWitt equation has tachyonic solutions since the mass-like term dependent on ${}^{(3)}R$ can be positive or negative.[165] A negative mass is an indication of tachyonic behavior wherein the wave propagation of the state functional is not necessarily in time-like directions and is thus tachyonic.

Eq. 10.3 is a Dirac-type D-dimensional Dirac equation. There are three other general types of universe particle equations. (By assumption fermionic universes come in four species like fermions.) The derivation of the four types of universe particles is similar to the derivation of fermion types in the Extended Standard Model in 4-dimensional complex space-time given in Blaha (2010b). We will now consider the D-dimensional equivalent for universe particles in the Megaverse.

The general form of a pure D-dimensional complex Lorentz group[166] boost can be expressed in terms of a complex relative (D – 1)-velocity $\mathbf{v_c}$ between inertial reference frames. A D-dimensional coordinate boost has the form

$$\Lambda_C(\mathbf{v_c}) \equiv \Lambda_C(\omega, \mathbf{v_c}) = \exp[i\omega\hat{\mathbf{w}}\cdot\mathbf{K}] \qquad (10.6)$$

where

$$\omega = (\omega_r^{\,2} - \omega_i^{\,2} + 2i\omega_r\omega_i\, \hat{\mathbf{u}}_r\cdot\hat{\mathbf{u}}_i)^{1/2} \qquad (10.7)$$

---

[164] The energy of universe particles need not be real-valued since universes can 'decay' – unlike elementary particles which are not subject to decay, by definition, since they are assumed to be *fundamental*. We choose to consider the case of universes with real-valued energies. The case of universes with complex-valued energies is a simple extension of the real-value cases considered here.

[165] DeWitt, B. S., Phys. Rev. **160**, 1113 (1967) p. 1124.

[166] The D-dimensional complex Lorentz group has similar features to the 4-dimensional complex Lorentz group. We shall only discuss it to the extent needed for our universe particle type's derivation. See Weinberg (1995) for the 4-dimensional Lorentz group – the D-dimensional Lorentz group generalizes directly from the features of the 4-dimensional Lorentz group.

and

$$\hat{\mathbf{w}} = (\omega_r\hat{\mathbf{u}}_r + i\omega_i\hat{\mathbf{u}}_i)/\omega \qquad (10.8)$$

with all vectors being (D – 1)-dimensional spatial vectors. We define the real and imaginary unit vectors $\hat{\mathbf{u}}_r\cdot\hat{\mathbf{u}}_r = 1 = \hat{\mathbf{u}}_i\cdot\hat{\mathbf{u}}_i$ with the result

$$\hat{\mathbf{w}}\cdot\hat{\mathbf{w}} = 1 \qquad (10.9)$$

The complex relative velocity is

$$\mathbf{v}_c = \hat{\mathbf{w}}\tanh(\omega) \qquad (10.10)$$

The free dynamical equations of the four universe particle species will be generated by D-dimensional Lorentz boosts of the free Dirac equation of a universe particle at rest with the *requirement that the time variable* ($t = y^D$) *and energy are real in the resulting field equations.*[167] The procedure can most easily be performed in D-dimensional momentum space with the Megaverse coordinate space version of the generated equation determined from the momentum space version.

10.3.1.1 Dirac-like Equation – Type I universe Particle

A positive energy plane wave solution of the Dirac equation eq. 10.3 for a universe particle at rest is

$$\psi(y) = \exp[-imt]w(0) \qquad (10.11)$$

where we set $\partial_t = \partial/\partial y_D$ while temporarily ignoring the $Y_u^{\,i}(y)/M_u^{D/2}$ term. w(0) is a $2^{D/2}$ component spinor column vector. The solution ψ(y) satisfies the momentum space Dirac equation for a particle at rest:

$$(m\gamma^D - m)\psi(y) = 0 \qquad (10.12)$$

The $2^{D/2} \times 2^{D/2}$ spinor matrix form of a D-dimensional Lorentz boost with a relative real velocity **v** of the Dirac matrices is[168]

$$S^{-1}(\Lambda(\mathbf{v}))\gamma^\nu S(\Lambda(\mathbf{v})) = \Lambda^\nu{}_\mu(\mathbf{v})\gamma^\mu \qquad (10.13)$$

where $\Lambda^\nu{}_\mu(\mathbf{v})$ is a D-dimensional Lorentz boost. S(Λ(**v**)) has the form

$$S(\Lambda(\mathbf{v})) = \exp(-\omega\gamma^D\boldsymbol{\gamma}\cdot\mathbf{v}/(2|\mathbf{v}|))$$

$$= \cosh(\omega/2)I + \sinh(\omega/2)\gamma^D\boldsymbol{\gamma}\cdot\mathbf{p}/|\mathbf{p}| \qquad (10.14)$$

---

[167] The D-dimensional "energy" must be real since it relates to the area of the universe – a real number.

[168] **The indices ν and μ from this point in this chapter have values: 1, 2, … , D.**

with *real* ω = arctanh(|**v**|) and *real* **v.** |**p**| is the magnitude of the spatial (D – 1)-vector. Also

$$S^{-1}(\Lambda(\mathbf{v})) = \gamma^D S^\dagger(\Lambda(\mathbf{v}))\gamma^D = \exp(\omega\gamma^D\boldsymbol{\gamma}\cdot\mathbf{v}/(2|\mathbf{v}|))$$

$$= \cosh(\omega/2)I - \sinh(\omega/2)\gamma^D\boldsymbol{\gamma}\cdot\mathbf{p}/|\mathbf{p}| \qquad (10.15)$$

If we now apply $S(\Lambda(\mathbf{v}))$ to the momentum space Dirac equation of a particle at rest (eq. 10.12) we find

$$0 = S(\Lambda(\mathbf{v}))(m\gamma^D - m)\ \psi(y)$$
$$= [mS(\Lambda(\mathbf{v}))\gamma^D S^{-1}(\Lambda(\mathbf{v})) - m]S(\Lambda(\mathbf{v}))w(0)$$

A straightforward evaluation shows

$$mS(\Lambda(v))\gamma^D S^{-1}(\Lambda(v)) = g_{D\mu\nu}p^\mu\gamma^\nu = \not{p} \qquad (10.16)$$

where p is a momentum D-vector. In addition we define the D-dimension spinor ($2^{D/2}$ components)

$$S(\Lambda(v))w(0) = w(p) \qquad (10.17)$$

which can be viewed as a "positive energy D Dirac spinor". The Dirac equation in momentum space has the familiar form:

$$(\not{p} - m)\exp[-ip\cdot y]w(p) = 0 \qquad (10.18)$$

Eq. 10.18 implies the free, coordinate space Dirac equation:

$$(i\gamma^\mu\partial/\partial y^\mu - m)\psi(y) = 0 \qquad (10.19)$$

We identify this equation as the dynamical equation of a type 1 universe particle. It corresponds to the free charged lepton elementary particle species Dirac equation in particle physics.

10.3.1.2 Complex Boosts

The form of the D-dimensional spinor boost transformation corresponding to the coordinate transformation eq. 10.6 is:

$$S_C(\omega, \mathbf{v_c}) \equiv S_C = \exp(-\omega\gamma^D\boldsymbol{\gamma}\cdot\hat{\mathbf{w}}/2)$$
$$= \cosh(\omega/2)I + \sinh(\omega/2)\gamma^D\boldsymbol{\gamma}\cdot\hat{\mathbf{w}} \qquad (10.20)$$

with *complex* $\mathbf{v_c}$ and $\hat{\mathbf{w}}$ defined by eqs. 10.10 and 10.8 respectively. The inverse transformation is

$$S_C^{-1}(\omega, \mathbf{v_c}) = \exp(\omega\gamma^D\boldsymbol{\gamma}\cdot\hat{\mathbf{w}}/2)$$

$$= \cosh(\omega/2)I - \sinh(\omega/2)\gamma^D\boldsymbol{\gamma}\cdot\hat{\mathbf{w}} \qquad (10.21)$$

Note that S_C is not unitary just as in the 4-dimensional case.

We now apply a spinor boost to the Dirac equation for a particle at rest in this more general case of complex ω and $\hat{\mathbf{w}}$.

$$0 = S_C(\omega, \mathbf{v_c}))(m\gamma^D - m)\exp[-imt]w(0)$$
$$= [mS_C\gamma^D S_C^{-1} - m]\exp[-imt]S_C w(0) \qquad (10.22)$$

where $S_C = S_C(\omega, \mathbf{v_c})$. After some algebra we find

$$mS_C\gamma^D S_C^{-1} = m[\cosh(\omega)\gamma^D - \sinh(\omega)\boldsymbol{\gamma}\cdot\hat{\mathbf{w}}] \qquad (10.23)$$

We will use these *complex* boosts to generate the other species' Dirac-like equations.

10.3.1.3 Tachyon Universe particle Dirac Equation

The development of the complex spinor boost transformation (subsection 10.3.1.2 above) leads to two possible forms of the tachyon Dirac-like equation. One form will lead to a lagrangian dynamics for left-handed universe particles. The other form leads to a lagrangian dynamics for right-handed universe particles.

10.3.1.4 Type IIa Case: Left-Handed Tachyonic Universe Particles

If the real and imaginary relative vectors parts of $\hat{\mathbf{w}}$, namely $\hat{\mathbf{u}}_r$ and $\hat{\mathbf{u}}_i$, are parallel, then $\hat{\mathbf{u}}_r\cdot\hat{\mathbf{u}}_i = 1$ and

$$\omega = \omega_r + i\omega_i \qquad (10.24)$$

Eqs. 10.23 and 10.24 then imply

$$mS_C\gamma^D S_C^{-1} = m[\cosh(\omega_r)\cos(\omega_i) + i\sinh(\omega_r)\sin(\omega_i)]\gamma^D -$$
$$- m[\sinh(\omega_r)\cos(\omega_i) + i\cosh(\omega_r)\sin(\omega_i)]\boldsymbol{\gamma}\cdot\hat{\mathbf{u}}_r \qquad (10.25)$$

or

$$mS_C\gamma^D S_C^{-1} = \cos(\omega_i)\gamma\cdot p_r + i\sin(\omega_i)\gamma\cdot p_i \qquad (10.26)$$

where

$$p_r^0 = m\cosh(\omega_r) \qquad p_i^0 = m\sinh(\omega_r) \qquad (10.27)$$

and

$$\mathbf{p_r} = m\hat{\mathbf{u}}_r\sinh(\omega_r) \qquad \mathbf{p_i} = m\hat{\mathbf{u}}_r\cosh(\omega_r) \qquad (10.28)$$

If $\omega_i = 0$, then we recover the momentum space Dirac-like equation. If $\omega_i = \pi/2$, then we obtain the left-handed momentum space tachyon equation:

$$mS_C\gamma^D S_C^{-1} = i\gamma{\cdot}p_i \tag{10.29}$$

and the tachyon energy and momentum expressions

$$\mathbf{p} = m\mathbf{v}\gamma_s \qquad\qquad E = m\gamma_s \tag{10.30}$$

where $\sinh(\omega) = \gamma_s = (\beta^2 - 1)^{-1/2}$ with $\beta = v/c > 1$. v is the absolute value of the (D – 1) component spatial velocity. Also

$$S_C w(0) = w_C(p) \tag{10.31}$$

is a tachyon spinor.

The momentum space tachyonic Dirac-like equation is

$$(i\not{p} - m)\exp[-ip{\cdot}y]w_T(p) = 0 \tag{10.32}$$

where $p{\cdot}y = p^D y^D - \mathbf{p{\cdot}y}$ after performing a corresponding boost in the exponential factor. If we apply $i\not{p}$ to eq. 10.32 we find the tachyon mass condition is satisfied

$$-E^2 + \mathbf{p}^2 = m^2 \tag{10.33}$$

Transforming back to coordinate space we obtain the "left-handed" *tachyonic Dirac-like equation*:

$$(\gamma^\mu\partial/\partial y^\mu - m)\psi_T(y) = 0 \tag{10.34}$$

10.3.1.5 Type IIb Case: Right-Handed Tachyonic Universe Particles

If the real and imaginary relative vectors parts of $\mathbf{\hat{w}}$, $\mathbf{\hat{u}_r}$ and $\mathbf{\hat{u}_i}$, are anti-parallel $\mathbf{\hat{u}_r} = -\mathbf{\hat{u}_i}$, then $\mathbf{\hat{u}_r}{\cdot}\mathbf{\hat{u}_i} = -1$ and

$$\omega = \omega_r - i\omega_i \tag{10.35}$$

then

$$mS_C\gamma^D S_C^{-1} = m[\cosh(\omega_r)\cos(\omega_i) - i\sinh(\omega_r)\sin(\omega_i)]\gamma^D - \\ - m[\sinh(\omega_r)\cos(\omega_i) - i\cosh(\omega_r)\sin(\omega_i)]\boldsymbol{\gamma}{\cdot}\mathbf{\hat{u}_r} \tag{10.36}$$

or

$$mS_C\gamma^D S_C^{-1} = \cos(\omega_i)\gamma{\cdot}p_r - i\sin(\omega_i)\gamma{\cdot}p_i \tag{10.37}$$

where

$$p_r^D = m\cosh(\omega_r) \qquad p_i^D = m\sinh(\omega_r) \tag{10.38}$$

and

$$\mathbf{p_r} = m\mathbf{\hat{u}_r}\sinh(\omega_r) \qquad\qquad \mathbf{p_i} = m\mathbf{\hat{u}_r}\cosh(\omega_r) \tag{10.39}$$

If $\omega_i = \pi/2$, then we obtain the right-handed momentum space tachyon equation.[169]

$$(-\gamma^\mu\partial/\partial y^\mu - m)\psi_T(y) = 0 \tag{10.40}$$

10.3.1.6 Type III Case: "Up-Quark-like" Universe Particles

There are two other cases where we can obtain fermion dynamical equations with a *real* time variable and real energy.[170] In one case we set $\hat{\mathbf{u}}_r\cdot\hat{\mathbf{u}}_i = 0$ and have a real ω.

If the real and imaginary relative vectors parts of $\hat{\mathbf{w}}$, namely $\hat{\mathbf{u}}_r$ and $\hat{\mathbf{u}}_i$, are perpendicular, $\hat{\mathbf{u}}_r\cdot\hat{\mathbf{u}}_i = 0$, then

$$\omega = (\omega_r^2 - \omega_i^2)^{1/2} \tag{10.41}$$

Thus ω is either pure real ($\omega_r \geq \omega_i$) or pure imaginary ($\omega_r < \omega_i$).

The momentum space equation generated by the corresponding spinor boost is

$$\{m \cosh(\omega)\gamma^D - m \sinh(\omega)\boldsymbol{\gamma}\cdot(\omega_r\hat{\mathbf{u}}_r + i\omega_i\hat{\mathbf{u}}_i)/\omega - m\} \exp[-imt]w_c(p) = 0 \tag{10.42}$$

Defining the momentum 4-vector

$$p = (p^D, \mathbf{p}) \tag{10.43}$$

where

$$p^D = m\cosh(\omega) \qquad \mathbf{p} = \mathbf{p}_r + i\mathbf{p}_i \tag{10.44}$$

with

$$\mathbf{p}_r = m\omega_r\hat{\mathbf{u}}_r \sinh(\omega)/\omega \qquad \mathbf{p}_i = m\omega_i\hat{\mathbf{u}}_i \sinh(\omega)/\omega \tag{10.45}$$

$$\mathbf{p}_r\cdot\mathbf{p}_i = 0 \tag{10.46}$$

then we obtain a positive energy Dirac-like equation

$$[p\cdot\gamma - m]\exp[-imt]w_c(p) = 0$$

or

$$[p^D\gamma^D - (\mathbf{p}_r + i\mathbf{p}_i)\cdot\boldsymbol{\gamma} - m]\exp[-ip\cdot y]w_c(p) = 0 \tag{10.47}$$

with a complex 3-momentum $\mathbf{p}$ and the 4-momentum mass shell condition:

$$p^2 = (p^D)^2 - \mathbf{p}_r\cdot\mathbf{p}_r + \mathbf{p}_i\cdot\mathbf{p}_i = m^2 \tag{10.48}$$

Note

---

[169] We note that $\gamma_s = (\beta^2 - 1)^{-1/2}$, *if expressed in terms of* ω*, has a branch cut extending from* $<-\infty, +\infty>$ *in the complex* ω *plane. Thus values of* ω *with positive imaginary parts are physically different from values of* ω *with negative imaginary parts.*

[170] The requirement of a real energy for a universe is not strict. For a fundamental free particle the energy must be real or the particle would be subject to decay – contrary to its assumed fundamental nature. Universes can 'decay' to 'smaller' universes. Therefore the requirement for real energy can be violated. Nevertheless the requirement for real energy is appealing since it leads to four species of universes strengthening the analogy of universes to elementary particles.

$$|\mathbf{v}_c| = |\mathbf{p}|/p^D = [(\mathbf{p}_r + i\mathbf{p}_i)\cdot(\mathbf{p}_r + i\mathbf{p}_i)]^{1/2}/p^D = \tanh(\omega) \quad (10.49)$$

and so the Lorentz factor is

$$\gamma = \cosh(\omega) \quad (10.50)$$

Eq. 10.47 is the momentum space equivalent of the wave equation[171]

$$[i\gamma^D\partial/\partial t + i\gamma\cdot(\nabla_r + i\nabla_i) - m]\psi_u(t, \mathbf{y}_r, \mathbf{y}_i) = 0 \quad (10.51)$$

where $\mathbf{y} = \mathbf{y}_r - i\mathbf{y}_i$, and where the grad operators ∇_r and ∇_i are with respect to $\mathbf{y}_r$ and $\mathbf{y}_i$ respectively. Since $\hat{\mathbf{u}}_r\cdot\hat{\mathbf{u}}_i = 0$ we see that there is a subsidiary condition on the wave function

$$\nabla_r\cdot\nabla_i\, \psi_u(t, \mathbf{y}_r, \mathbf{y}_i) = 0 \quad (10.52)$$

We note eq. 10.52 can be put into covariant form as the difference of two vectors squared (which is a real D-dimensional Lorentz group invariant):

$$[\gamma^D\partial/\partial t + i\gamma\cdot(\nabla_r + i\nabla_i)]^2 - [\gamma^D\partial/\partial t + i\gamma\cdot(\nabla_r - i\nabla_i)]^2 = 4\nabla_r\cdot\nabla_i.$$

We identify eq. 10.51 as the dynamical equation of an "up-quark-like" universe particle.

10.3.1.7 Type IVa Case: Left-Handed "Down-Quark-like" Tachyonic Universe Particles

In this case we set $\hat{\mathbf{u}}_r\cdot\hat{\mathbf{u}}_i = 0$. Then by eq. 10.7

$$\omega = (\omega_r^2 - \omega_i^2)^{1/2}$$

Thus ω again starts out either pure real (if $\omega_r \geq \omega_i$) or pure imaginary (if $\omega_r < \omega_i$). In this case we also choose ω real, and then change ω to

$$\omega = (\omega_r^2 - \omega_i^2)^{1/2} \rightarrow \omega' = (\omega_r^2 - \omega_i^2)^{1/2} + i\pi/2 = \omega + i\pi/2$$

by adding $i\pi/2$ to ω since ω is a free parameter. We then proceed as we did in the prior tachyon case.[172]. The resulting Lorentz boost

---

[171] The gradient operators ∇_r and ∇_i are 15-dimensional spatial gradient operators.

[172] Here again the choice of ω in eq. 10.53 leads to a "left-handed" universe particle while the choice $\omega' = \omega - i\pi/2$ leads to a right-handed one.

$$\Lambda_C = \exp[i((\omega_r^2 - \omega_i^2)^{1/2} + i\pi/2)(\omega_r\hat{\mathbf{u}}_r + i\omega_i\hat{\mathbf{u}}_i)\cdot\mathbf{K}/\omega] \qquad (10.53)$$

becomes a left-handed “quark-like” boost. The tachyon dynamical equation is[173]

$$[\gamma^D\partial/\partial t + \boldsymbol{\gamma}\cdot(\nabla_r + i\nabla_i) - m]\psi_d(y) = 0 \qquad (10.54)$$

with the constraint equation

$$\nabla_r\cdot\nabla_i\ \psi_d(t, \mathbf{y}_r, \mathbf{y}_i) = 0 \qquad (10.55)$$

We will call the universe particles satisfying eqs. 10.54 and 10.55 left-handed *tachyonic quark-like universe particles.*

10.3.1.8 Type IVb Case: Right-Handed Down-Quark-like Tachyonic Universe Particles

In this case we set $\hat{\mathbf{u}}_r\cdot\hat{\mathbf{u}}_i = 0$. Then by eq. 10.7

$$\omega = (\omega_r^2 - \omega_i^2)^{1/2}$$

Thus ω again starts out either pure real (if $\omega_r \geq \omega_i$) or pure imaginary (if $\omega_r < \omega_i$). In this case we also choose ω real, and then change ω to

$$\omega = (\omega_r^2 - \omega_i^2)^{1/2} \rightarrow \omega' = (\omega_r^2 - \omega_i^2)^{1/2} - i\pi/2 = \omega - i\pi/2$$

since ω is a free parameter and proceed as we did in the prior case. The resulting Lorentz boost

$$\Lambda_C = \exp[i((\omega_r^2 - \omega_i^2)^{1/2} - i\pi/2)(\omega_r\hat{\mathbf{u}}_r + i\omega_i\hat{\mathbf{u}}_i)\cdot\mathbf{K}/\omega] \qquad (10.56)$$

becomes a right-handed quark-like boost. The resulting tachyon dynamical equation is

$$[-\gamma^D\partial/\partial t - \boldsymbol{\gamma}\cdot(\nabla_r + i\nabla_i) - m]\psi_d(y) = 0 \qquad (10.57)$$

with the constraint equation

$$\nabla_r\cdot\nabla_i\ \psi_d(t, \mathbf{y}_r, \mathbf{y}_i) = 0 \qquad (10.58)$$

We will call the universe particles satisfying eqs. 10.57 and 10.58 right-handed *tachyonic quark-like universe particles.*

---

[173] The gradient operators ∇_r and ∇_i are (D – 1)-dimensional spatial gradient operators.

10.3.2 Lagrangians

In this section we will develop a lagrangian formalism for each of the four types of fermionic universe particles noting that a tachyonic universe particles have two forms: left-handed and right-handed (discussed later in section 10.3.5).

The various types of universe particles described in section 10.3.1 correspond to universes with differing internal characteristics and motion in the Megaverse. The equations are all free field equations. Internal potentials and interactions must be introduced in these equations to complete the universe dynamical equations. A connection to the Wheeler-DeWitt description of their internal quantum structure also remains to be established (section 10.3.6).

In defining the lagrangians for the four fermionic universe types that yield their dynamical equations in a canonical manner, we require the conventional quantum field theory feature that the hamiltonian derived from the lagrangian is hermitean. We will develop a separate lagrangian for each type.

10.3.2.1 Type I Universe Particle Lagrangian

The Universe particle Dirac equation lagrangian is

$$\mathcal{L}_u = \bar{\psi}(i\gamma^\mu \partial/\partial y^\mu - m)\psi(y) \tag{10.59}$$

where

$$\bar{\psi} = \psi^\dagger \gamma^D$$

and $\psi^\dagger$ is the hermitean conjugate of ψ.

10.3.2.2 Type II Tachyon Universe Particle Lagrangian

This lagrangian includes both left-handed and right-handed cases. It can be separated into lagrangian terms for each case using parity projection operators.

$$\mathcal{L}_{uT} = \psi_T{}^S(\gamma^\mu \partial/\partial y^\mu - m)\psi_T(y) \tag{10.60}$$

where

$$\psi_T{}^S = \psi_T{}^\dagger\, i\gamma^D\gamma^5 \tag{10.61}$$

with γ^5 being the D-dimensional equivalent for γ^5 in 4 dimensions. The peculiar form of the tachyon universe lagrangian is necessitated by the hermiticity of the hamiltonian calculated from it.

10.3.2.3 Type III "Up-Quark-like" Universe Particle Lagrangian

The lagrangian density of a free "up-quark-like" universe particle is

$$\mathcal{L}_u = \bar{\psi}_u(i\gamma^\mu D_\mu - m)\psi_u(y) \tag{10.62}$$

where $\bar{\psi}_u = \psi_u{}^\dagger \gamma^D$ and

$$\psi_u^{\dagger} = [\psi_u(\mathbf{y_r}, \mathbf{y_i})]^{\dagger} |_{\mathbf{y_i} = -\mathbf{y_i}} \quad (10.63)$$

$$D_D = \partial/\partial y^D$$
$$D_k = \partial/\partial y_r^{\,k} + i\,\partial/\partial y_i^{\,k} \quad (10.64)$$

for k = 1, 2, … , (D – 1). The action

$$I = \int d^{(D-1)}y\,\mathcal{L}_u \quad (10.65)$$

It is easy to show that this action is also real.

10.3.2.4 Type IV "Down-Quark-like" Tachyon Universe Particle Lagrangian

The lagrangian density of a free "down-quark-like" universe particle is

$$\mathcal{L}_d = \psi_d^{\,C}(y)(\gamma^{\mathbf{D}}\partial/\partial t + \boldsymbol{\gamma}\cdot(\nabla_{\mathbf{r}} + i\nabla_{\mathbf{i}}) - m)\psi_d(y) \quad (10.66)$$

where

$$\psi_d^{\,C}(y) = [\psi_d(y)]^{\dagger}|_{\mathbf{y_i} = -\mathbf{y_i}}\, i\gamma^D\gamma^5 \quad (10.67)$$

In words, eq. 10.67 states: take the hermitean conjugate of $\psi_d(y)$; change $\mathbf{y_i}$ to $-\mathbf{y_i}$; and then post-multiply by the indicated factors.

The action is

$$I = \int d^{(D-1)}y\,\mathcal{L}_d \quad (10.68)$$

The action is real. The lagrangian can also be separated into left-handed and right-handed parts using projection operators.

10.3.3 Form of The Megaverse Quantum Coordinates Gauge Field

The discussions of sections 10.3.1 and 10.3.2 assumed the coordinates were Megaverse coordinates and their derivatives. Prior to those discussions we indicated we would use quantum coordinates in the Megaverse of the form[174]

$$Y^i(y) = y^i + i\,Y_u^{\,i}(y)/M_u^{\,D/2} \quad (10.4)$$

and their derivatives

$$\partial_i = \partial/\partial Y^i(y) = \partial/\partial(y^i - Y_u^{\,i}(y)/M_u^{\,D/2}) \quad (10.5)$$

for i = 1, 2, … , D to eliminate divergences in quantum field theory. The subscript "u" signifies universes. The mass constant for the Megaverse, M_u, may be the same as the

[174] The denominator $M_u^{\,D/2}$ is necessitated by the dimension of $Y_u^{\,i}(y)$ which is $[m]^{D/2-1}$. Eqs. 10.78 and 10.81 below imply this conclusion.

mass constant M_c appearing in the Two Tier mechanism for our universe. (See chapter 7 for a discussion of eliminating infinities with this mechanism.)

In this section we define the gauge fields $Y_u^i(y)$ of the Megaverse.[175] They are similar to the $Y^\mu(y)$ fields of our New Standard Model.[176] The $Y_u(y)$ D-dimensional vector gauge field, in the absence of external sources, will be defined in a D-dimensional Coulomb gauge:

$$Y_u^D(y) = 0 \qquad (10.69)$$
$$\partial Y_u^j(y)/\partial y^j = 0$$

where the sum over j is over the D – 1 spatial y coordinates. We follow a procedure similar to Blaha (2003) but for D-dimensional space. The lagrangian density for the free $Y_u^j(y)$ fields is

$$\mathscr{L}_u = -\tfrac{1}{4} F_u^{\mu\nu} F_{u\mu\nu} \qquad (10.70)$$

and the lagrangian is

$$L_u = \int d^{(D-1)}y \mathscr{L}_u \qquad (10.71a)$$

with

$$F_{u\mu\nu} = \partial Y_{u\mu}/\partial y^\nu - \partial Y_{u\nu}/\partial y^\mu \qquad (10.71b)$$

The equal time commutation relations, derived in the usual way, are:

$$[Y_u^\mu(\mathbf{y}, y^0), Y_u^\nu(\mathbf{y'}, y^0)] = [\pi_u^\mu(\mathbf{y}, y^0), \pi_u^\nu(\mathbf{y'}, y^0)] = 0 \qquad (10.72)$$
$$[\pi_u^j(\mathbf{y}, y^0), Y_{uk}(\mathbf{y'}, y^0)] = -i\, \delta^{(D-1)tr}{}_{jk}(\mathbf{y} - \mathbf{y'}) \qquad (10.73)$$

for μ, ν, j, k = 1, 2, … , (D – 1) where

$$\pi_u^k = \partial \mathscr{L}_u / \partial Y_{uk}' \qquad (10.74)$$
$$\pi_u^0 = 0 \qquad (10.75)$$

and

$$\delta^{tr}{}_{jk}(\mathbf{y} - \mathbf{y'}) = \int d^{(D-1)}k\, e^{i\,\mathbf{k}\bullet(\mathbf{y}-\mathbf{y'})} (\delta_{jk} - k_j k_k/\mathbf{k}^2)/(2\pi)^{D-1} \qquad (10.76)$$

$$Y_{uk}' = \partial Y_{uk}/\partial y^D \qquad (10.77)$$

The Coulomb gauge indicates D – 2 degrees of freedom are present in the vector potential. The Fourier expansion of the vector potential is:

$$Y_u^i(y) = \int d^{(D-1)}k\, N_0(k) \sum_{\lambda=1}^{D-2} \varepsilon^i(k, \lambda)[a(k,\lambda)\, e^{-ik\cdot y} + a^\dagger(k,\lambda)\, e^{ik\cdot y}] \qquad (10.78)$$

---

[175] This choice implies that the Megaverse Y mass $M_u = M_C$, its universe mass.
[176] See Blaha (2005a) for details.

where

$$N_0(k) = [(2\pi)^{(D-1)} 2\omega_k]^{-1/2} \quad (10.79)$$

and (since the field is massless)

$$k^D = \omega_k = (\mathbf{k}^2)^{1/2} \quad (10.80)$$

where k^D is the energy, and where the $\varepsilon^i(k, \lambda)$ are the polarization unit vectors for $\lambda = 1, \ldots, (D-2)$ and $k^\mu k_\mu = k^{D\,2} - \mathbf{k}^2 = 0$.

The commutation relations of the Fourier coefficient operators are:

$$[a(k,\lambda), a^\dagger(k',\lambda')] = \delta_{\lambda\lambda'}\delta^{(D-1)}(\mathbf{k} - \mathbf{k}') \quad (10.81)$$
$$[a^\dagger(k,\lambda), a^\dagger(k',\lambda')] = [a(k,\lambda), a(k',\lambda')] = 0 \quad (10.82)$$

and the polarization vectors satisfy

$$\sum_{\lambda=1}^{D-2} \varepsilon_i(k, \lambda)\varepsilon_j(k, \lambda) = (\delta_{ij} - k_i k_j/\mathbf{k}^2) \quad (10.83)$$

It will be convenient to divide the Y field into positive and negative frequency parts:

$$Y_u{}^+{}_i(y) = \int d^{(D-1)}k\, N_0(k) \sum_{\lambda=1}^{D-2} \varepsilon_i(k, \lambda)\, a(k,\lambda)\, e^{-ik\cdot y} \quad (10.84)$$

and

$$Y_u{}^-{}_i(y) = \int d^{(D-1)}k\, N_0(k) \sum_{\lambda=1}^{D-2} \varepsilon_i(k, \lambda)\, a^\dagger(k,\lambda)\, e^{ik\cdot y} \quad (10.85)$$

For later use we note the commutator between the positive and negative frequency parts is:

$$[\, Y_u{}^-{}_j(y_1), Y_u{}^+{}_k(y_2)] = -\int d^{(D-1)}k\, e^{ik\cdot(y_1 - y_2)}\, (\delta_{jk} - k_j k_k/\mathbf{k}^2)/[(2\pi)^{D-1} 2\omega_k] \quad (10.86)$$

10.3.3.1 Y^μ Fock Space Imaginary Coordinate States

States can also be defines for the quantized Y^μ field. These states will be similar in form to electromagnetic photon states but play a different role in our approach since they are in fact coordinate excitation states for the imaginary part of $Y^i(y)$ (eq. 10.4). Thus universe particles (and other fields) will exist in a real D-dimensional space with quantum excitations into imaginary Quantum Dimensions. These excitations become significant at high energies. At low energies space appears as c-number complex; at very high energies space becomes slightly q-number complex.

There are two types of imaginary coordinate excitations: 1.) Quantum excitations into Fock states consisting of a superposition of states with a definite finite number of Y_u "particles" and 2.) Imaginary coordinate excitations into coherent Y_u states with an "infinite" number of particles. Coherent states can be viewed as representing "classical" fields.

In this section we will consider Y_u field states with a definite number of excitations ("particles"). The raising and lowering operators of the Y_u field can be used to define free particle states. For example a one particle state can be defined by

$$|k, \lambda> = a^{\dagger}(k, \lambda)|0> \qquad (10.87)$$

with corresponding bra state

$$<k, \lambda| = <0|a(k, \lambda) \qquad (10.88)$$

where the "coordinate vacuum" is defined as usual:

$$a(k, \lambda)|0> = 0 \qquad (10.89)$$

$$<0|a^{\dagger}(k, \lambda) = 0 \qquad (10.90)$$

Multi-particle states can also be defined in the conventional way with products of the raising and lowering operators applied to the vacuum. The set of all states containing a finite number of "particles" constitutes a Fock space.

A state with a finite number of Y_u "particles" represents a quantum fluctuation into imaginary Quantum Dimensions.

10.3.3.2 Y_u Coherent Imaginary Coordinate States

Coherent Y_u states bring us closer what we might consider to be "classical" imaginary dimensions – dimensions that we can, in principle, experience as we do normal dimensions. Let us define the coherent state[177]

$$| y, p> = e^{-\mathbf{p}\cdot Y_u^-(y)/M_u^{D/2}}|0> \qquad (10.91)$$

This state is an eigenstate of the coordinate operator $Y_u^+(y')$:

$$Y_u{}^+{}_j(y_1)\,|y_2, p> = -[Y_u{}^+{}_j(y_1), \mathbf{p}\cdot\mathbf{Y}^-(y_2)]/M_u^{D/2}|y, p> \qquad (10.92)$$

$$= -\int d^{D-1}k\,[N_0(k)]^2\, e^{ik\cdot(y_2-y_1)}\,(p_j - k_j\mathbf{p}\cdot\mathbf{k}/\mathbf{k}^2)/M_u^{D/2}|y, p>$$
$$= p^i\Delta_{Tij}(y_1 - y_2)/M_u^{D/2}|y, p> \qquad (10.93)$$

[177] Coherent states are well known in the physics literature. See for example T. W. B. Kibble, J. Math. Phys. **9**, 315 (1968) and references therein; V. Chung, Phys. Rev. **140**, B1110 (1965); J. R. Klauder, J. McKenna, and E. J. Woods, J. Math. Phys. **7**, 822 (1966) and references therein.

where $p^i\Delta_{Tij}(y_1 - y_2)/M_u^{D/2}$ is the eigenvalue of $Y_u{}^+{}_j(y_1)$. As we will see later, the eigenvalue of $Y_u{}^+$ becomes large as $(y_1 - y_2)^2 \rightarrow 0$. Thus the imaginary Quantum Dimensions become significant at very short distances, and then significantly modifies the high-energy behavior of quantum field theories. In particular, Quantum Dimensions have a significant effect when

$$(y_1 - y_2)^2 \lessapprox (2^{D-2}\pi^{D-2}M_u^2)^{-1} \tag{10.94}$$

We assume the mass scale $M_u = M_C$ is very large – perhaps of the order of the Planck mass (1.221×10^{19} GeV/c^2).

10.3.3.3 Quantization of the Type I Free Universe Particle Dirac Field

The quantization procedure is formally identical to that of a conventional Dirac particle. The standard equal time anti-commutation relations for a D-dimensional fermion field are:

$$\{\psi_\alpha(Y), \psi_\beta(Y')\} = \{\pi_{\psi\alpha}(Y), \pi_{\psi\beta}(Y')\} = 0 \tag{10.95}$$
$$\{\pi_{\psi\alpha}(Y), \psi_\beta(Y')\} = i\,\delta_{\alpha\beta}\,\delta^{D-1}(\mathbf{Y} - \mathbf{Y'}) \tag{10.96}$$

where α and β are the spinor indices ranging from 1 to $N_{MRC} = 2^{D/2}$ and where

$$\pi_{\psi\alpha}(Y) = i\,\psi_\alpha{}^\dagger(Y) \tag{10.97}$$

The field can be expanded in a fourier series:

$$\psi(Y(y)) = \sum_s\int d^{D-1}p\, N^d{}_m(p)\, [b(p,s)u(p,s) :e^{-ip\cdot(y + iYu/M_u^{D/2})}: + d^\dagger(p,s)v(p,s) :\exp(ip\cdot(y + $$
$$+ iYu/Mu^{D/2})):] \tag{10.98}$$

$$\psi^\dagger(Y(y)) = \sum_s\int d^{D-1}p\, N^d{}_m(p)\, [b^\dagger(p,s)\bar{u}(p,s)\gamma^0 :e^{+ip\cdot(y + iYu/M_u^{D/2})}: + d(p,s)\bar{v}(p,s)\gamma^0 \cdot$$
$$\cdot :\exp(-ip\cdot(y + iYu/Mu^{D/2}):] \tag{10.99}$$

where

$$N^d{}_m(p) = [m/((2\pi)^{D-1}E_p)]^{1/2} \tag{10.100}$$

and

$$E_p = p^D = (\mathbf{p}^2 + m^2)^{1/2} \tag{10.101}$$

with : … : signifying normal ordering. The commutation relations of the Fourier coefficient operators are:

$$\{b(p,s), b^\dagger(p',s')\} = \delta_{ss'}\delta^{D-1}(\mathbf{p} - \mathbf{p'}) \tag{10.102}$$
$$\{d(p,s), d^\dagger(p',s')\} = \delta_{ss'}\delta^{D-1}(\mathbf{p} - \mathbf{p'}) \tag{10.103}$$

$$\{b(p,s), b(p',s')\} = \{d(p,s), d(p',s')\} = 0 \tag{10.104}$$
$$\{b^\dagger(p,s), b^\dagger(p',s')\} = \{d^\dagger(p,s), d^\dagger(p',s')\} = 0 \tag{10.105}$$
$$\{b(p,s), d^\dagger(p',s')\} = \{d(p,s), b^\dagger(p',s')\} = 0 \tag{10.106}$$
$$\{b^\dagger(p,s), d^\dagger(p',s')\} = \{d(p,s), b(p',s')\} = 0 \tag{10.107}$$

The spinors u(p,s) and v(p,s) are defined in a conventional way (as in Bjorken and Drell). However their form is different from the 4-dimensional case. If one takes the $N_{MRC} \times N_{MRC} \equiv 2^{D/2} \times 2^{D/2}$ $\gamma \cdot p$ matrix, then the first $2^{D/2-1}$ columns give u(p,s) up to a normalization for the free particle case, the remaining $2^{D/2-1}$ columns give v(p,s) up to a normalization.

10.3.3.4 Feynman Propagators for the Type I Free Universe Particle Dirac Field

The form of the fermionic universe particle Feynman propagator differs from a conventional fermion propagator by having a Gaussian factor R(**p**, z) in its fourier expansions. This follows from using quantum Megaverse coordinates (eq. 10.4).

$$iS_F^{TT}(y_1 - y_2) = <0|T(\bar{\psi}(Y(y_1))\psi(Y(y_2)))|0> \tag{10.108}$$

where the time ordering is with respect to y_1^D and y_2^D. Expanding the free fields leads to the fourier representation:

$$iS_F^{TT}(y_1 - y_2) = i \int \frac{d^D p \, e^{-ip\cdot(y_1 - y_2)} (\not{p} + m) R(\mathbf{p}, y_1 - y_2)}{(2\pi)^D (p^2 - m^2 + i\varepsilon)} \tag{10.109}$$

where

$$R(\mathbf{p}, y_1 - y_2) = \exp[-p^i p^j \Delta_{Tij}(y_1 - y_2)/M_u^D] \tag{10.110}$$
$$= \exp\{-p^2[A(v) + B(v)\cos^2\theta] / [(2\pi)^{D-2} M_c^4 z^2]\} \tag{10.111}$$

(Note p^2 is the square of the spatial (D – 1)-vector.) with

$$z^\mu = y_1^\mu - y_2^\mu \tag{10.112}$$
$$z = |\mathbf{z}| = |\mathbf{y_1} - \mathbf{y_2}| \tag{10.113}$$
$$p = |\mathbf{p}| \tag{10.114}$$
$$v = |z^0|/z \tag{10.115}$$
$$A(v) = (1 - v^2)^{-1} + .5v \ln[(v-1)/(v+1)] \tag{10.116}$$
$$B(v) = v^2(1 - v^2)^{-1} - 1.5v \ln[(v-1)/(v+1)] \tag{10.117}$$
$$\mathbf{p}\cdot\mathbf{z} = pz\cos\theta \tag{10.118}$$

and |**p**| denoting the length of a spatial (D – 1)-vector **p** while $|z^0|$ is the absolute value of $z^0 \equiv z^D$.

As eq. 10.109 indicates, the Gaussian damping factor[178] R(p, z) for large spatial momentum p is the same for both the positive and negative frequency parts of the Two Tier Feynman propagator. We are assuming the spatial momentum is real-valued in this discussion. It is also important to note that R(p, z) does not depend on $p^0 = p^D$ (in the Y Coulomb gauge) and thus the integration over p^0 proceeds in the usual way to produce time-ordered positive and negative frequency parts.

10.3.3.5 Feynman Propagators for the Types II, III, and IV Free Universe Particle Dirac Fields

These propagators differ in details from the Type I propagator. The differences modulo the change in dimension appear in Blaha (2011c). See also Blaha (2005a) for a detailed discussion of 4-dimensional spin ½ particle propagators.

10.3.4 Expanding and Contracting Universes: Impact of Time Dependent Universe Particle Masses

Our discussions of the dynamics of universe particles assumed their masses were constant. However the definition of mass in terms of the area of a universe based on the physics of black holes is

$$M = \kappa A/8\pi \tag{10.119}$$

where A is the area of the black hole shows that *the mass of a universe particle is time dependent* because the area of a universe is generally time dependent. For example, our universe is expanding and its surface area is thus growing with time.

Eqs. 10.11 (and subsequent fermionic dynamic equations) must then be modified from

$$\psi(y) = \exp[-imt]w(0) \tag{10.11}$$

to a covariant form:

$$\psi(y) = \exp[-i\int_0^{w\cdot y} m(t')dt']w(0) \tag{10.120}$$

where w is a unit D-vector in the time (y^D) direction ($w^2 = 1$). The lower bound on the integral, 0, is the time of the beginning of the universe particle – its Big Bang. Thus the cumulative change in the mass of the universe particle may be significant. It is interesting to note that the Wheeler-Dewitt equation also has a variable value mass term R that also depends on the evolution of the universe.

Eq. 10.120 satisfies the free covariant Dirac-like universe particle field dynamic equation

$$[i\gamma^i\partial/\partial y^i - m(w\cdot y)]\psi(y) = 0 \tag{10.121}$$

[178] Note the Gaussian damping is for all D – 1 spatial momentum integrations.

In contrast to the constant mass equation eq. 10.19. Substituting eq. 10.120 in eq. 10.121 we find

$$(\gamma^i w_i\, m(w\cdot y) - m(w\cdot y))\psi(y) = 0 \qquad (10.122)$$

or

$$(\gamma^i w_i - 1)\psi(y) = 0 \qquad (10.123)$$

Upon performing a D-dimensional Lorentz boost (of the type of eqs. 10.13 – 10.16) on eq. 10.123 we obtain

$$(\gamma_i p^i/m_0 - 1)\psi(y) = 0$$

or

$$(\gamma_i p^i - m_0)\psi(y) = 0 \qquad (10.124)$$

where p^i is a momentum D-vector with $p^2 = m_0^2$. Eq. 10.123 is the constant mass momentum space dynamic equation. It determines the spinor in $\psi(y)$. After taking account of the quantum coordinates the quantum Dirac-like universe particle wave function has the form

$$\psi(Y(y)) = \sum_s \int d^{(D-1)}p\, N^d{}_m(p)\, [b(p,s)u(p,s) : \exp[-iG(p, Y(y))]: + d^\dagger(p,s)v(p,s) \cdot$$
$$\cdot :\exp[+iG(p, Y(y))]:\} \qquad (10.125)$$

$$\psi^\dagger(Y(y)) = \sum_s \int d^{(D-1)}p\, N^d{}_m(p)\, \{b^\dagger(p,s)\bar{u}(p,s)\gamma^0 :\exp[+iG(p, Y(y))]: + d(p,s)\bar{v}(p,s)\gamma^0 \cdot$$
$$\cdot \exp[-iG(p, Y(y))]:\} \qquad (10.126)$$

where : … : denotes normal ordering and

$$G(p, Y(y)) = \int_0^{p\cdot Y(y)/\lambda} m(t')dt' \qquad (10.127)$$

with $\lambda = m_0$, and $N^d{}_m(p)$ a normalization constant. Contrast eqs. 10.125-10.126 to the constant mass case eqs. 10.98-10.101. The *constant mass case* simply sets $m(t') = m_0$.

If we examine the integral eq. 10.127 for a short time interval δt in the particle's rest frame then $G(p, Y(y)) \approx m(0)\delta t$ and so we define $m(0) = m_0$. Based on the formula for universe particle mass (eq. 10.119) we anticipate that m_0 might be as large as the Planck mass or larger – thus an extremely short radius. Blaha (2013) describes a quantum Big Bang model in which the initial radius of the universe is $O(EM_{Planck}^{-2})$ where E is of the order of 1 and has the dimensions of [mass].

Thus we have a closed form definition of a quantum universe particle wave function for universe particles of type I. A similar procedure can be followed for universe particles of types II, III, and IV.

The Feynman propagator for type I quantum fields is *not* eq. 10.109 but now has a form reflecting the Y(y) dependence of the quantum fields in eqs. 10.125 and 10.126:

$$iS_F^{TT}(y_1, y_2) = i \int \frac{d^Dp \ \{ <0|\theta(y_{1D} - y_{2D})G(y_1, y_2) + \theta(y_{2D} - y_{1D})G(y_2, y_1)\}0>}{(2\pi)^D (p - m_0)} \tag{10.128}$$

where p^D is the energy and

$$G(y_1, y_2) = :\exp[-iG(p, Y(y_1))]: :\exp[+iG(p, Y(y_2))]: \tag{10.129}$$

Let

$$\begin{aligned} G_{tot}(y_1, y_2) &= <0|\theta(y_{1D} - y_{2D})G(y_1, y_2) + \theta(y_{2D} - y_{1D})G(y_2, y_1)\}0> \quad (10.130) \\ &= <0|\theta(y_{1D} - y_{2D}):\exp[-iG(p, Y(y_1))]::\exp[+iG(p, Y(y_2))]: + \\ &\quad + \theta(y_{2D} - y_{1D}) :\exp[-iG(p, Y(y_2))]::\exp[+iG(p, Y(y_1)):]|0> \end{aligned}$$

$$= <0|\theta(y^D{}_1 - y^D{}_2): \exp[-i\int_0^{p\cdot Y(y1)/\lambda} m(t')dt']::\exp[+i\int_0^{p\cdot Y(y2)/\lambda} m(t')dt']: +$$

$$+ \theta(y^D{}_2 - y^D{}_1):\exp[-i \exp[+i\int_0^{p\cdot Y(y2)/\lambda} m(t')dt']::\exp[+i\int_0^{p\cdot Y(y1)/\lambda} m(t')dt']:|0>$$

with $\lambda = m_0$ then

$$iS_F^{TT}(y_1, y_2) = i \int \frac{d^Dp \ G_{tot}(y_1, y_2)}{(2\pi)^D (p - m_0)} \tag{10.131}$$

Except for the case of a constant mass, where $m(t) = m_0$, the Feynman propagator is not a function of $y_1 - y_2$. The evaluation of eq. 10.130 in the general case of a variable mass is straightforward but cumbersome. For the special case of a linear time dependence of the mass, $m(t) = at$, we find eq. 10.130 gives

$$\begin{aligned} G_{tot}(y_1, y_2) =&<0|\theta(y^D{}_1 - y^D{}_2):\exp[-ia(p\cdot Y(y_1)/m_0)^2/2]::\exp[+ia(p\cdot Y(y_2)/m_0)^2/2]: + \\ &+ \theta(y^D{}_1 - y^D{}_2):\exp[-ia(p\cdot Y(y_2)/m_0)^2/2]::\exp[+ia(p\cdot Y(y_1)/m_0)^2/2]:|0> \end{aligned} \tag{10.132}$$

yielding a complex function of p, y_1, and y_2. *Note that the lower bound of the integrals in the Feynman propagator cancel and thus the need for an understanding of the beginning of a universe is removed in this case.*

We have shown that universe particle theory can handle the case of a variable universe mass m(t). Expanding or contracting (or oscillating) universe particles correspond to expanding and contracting (or oscillating) universes.

REFERENCES

Akhiezer, N. I., Frink, A. H. (tr), 1962, *The Calculus of Variations* (Blaisdell Publishing, New York, 1962).

Bjorken, J. D., Drell, S. D., 1964, *Relativistic Quantum Mechanics* (McGraw-Hill, New York, 1965).

Bjorken, J. D., Drell, S. D., 1965, *Relativistic Quantum Fields* (McGraw-Hill, New York, 1965).

Blaha, S., *C++ for Professional Programming* (International Thomson Publishing, Boston, 1995).

_______, 1998, *Cosmos and Consciousness* (Pingree-Hill Publishing, Auburn, NH, 1998 and 2002).

_______, 2002, *A Finite Unified Quantum Field Theory of the Elementary Particle Standard Model and Quantum Gravity Based on New Quantum Dimensions™ & a New Paradigm in the Calculus of Variations* (Pingree-Hill Publishing, Auburn, NH, 2002).

_______, 2004, *Quantum Big Bang Cosmology: Complex Space-time General Relativity, Quantum Coordinates™ Dodecahedral Universe, Inflation, and New Spin 0, ½, 1 & 2 Tachyons & Imagyons* (Pingree-Hill Publishing, Auburn, NH, 2004).

______, *2005a, Quantum Theory of the Third Kind: A New Type of Divergence-free Quantum Field Theory Supporting a Unified Standard Model of Elementary Particles and Quantum Gravity based on a New Method in the Calculus of Variations* (Pingree-Hill Publishing, Auburn, NH, 2005).

______, 2005b, *The Metatheory of Physics Theories, and the Theory of Everything as a Quantum Computer Language* (Pingree-Hill Publishing, Auburn, NH, 2005).

______, 2005c, *The Equivalence of Elementary Particle Theories and Computer Languages: Quantum Computers, Turing Machines, Standard Model, Superstring Theory, and a Proof that Gödel's Theorem Implies Nature Must Be Quantum* (Pingree-Hill Publishing, Auburn, NH, 2005).

______, 2006a, *The Foundation of the Forces of Nature* (Pingree-Hill Publishing, Auburn, NH, 2006).

______, 2006b, *A Derivation of ElectroWeak Theory based on an Extension of Special Relativity; Black Hole Tachyons; & Tachyons of Any Spin.* (Pingree-Hill Publishing, Auburn, NH, 2006).

_______, 2007a, *Physics Beyond the Light Barrier: The Source of Parity Violation, Tachyons, and A Derivation of Standard Model Features* (Pingree-Hill Publishing, Auburn, NH, 2007).

______, 2007b, *The Origin of the Standard Model: The Genesis of Four Quark and Lepton Species, Parity Violation, the ElectroWeak Sector, Color SU(3), Three Visible Generations of Fermions, and One Generation of Dark Matter with Dark Energy* (Pingree-Hill Publishing, Auburn, NH, 2007).

______, *2008a, A Direct Derivation of the Form of the Standard Model From GL(16) (Pingree-Hill Publishing, Auburn, NH, 2008).*

______, 2008b, *A Complete Derivation of the Form of the Standard Model With a New Method to Generate Particle Masses Second Edition* (Pingree-Hill Publishing, Auburn, NH, 2008)

______, 2009, *The Algebra of Thought & Reality: The Mathematical Basis for Plato's Theory of Ideas, and Reality Extended to Include A Priori Observers and Space-Time Second Edition* (Pingree-Hill Publishing, Auburn, NH, 2009).

______, 2010a, *Operator Metaphysics: A New Metaphysics Based on a New Operator Logic and a New Quantum Operator Logic that Lead to a Mathematical Basis for Plato's Theory of Ideas and Reality* (Pingree-Hill Publishing, Auburn, NH, 2010).

______, 2010b, *The Standard Model's Form Derived from Operator Logic, Superluminal Transformations and GL(16)* (Pingree-Hill Publishing, Auburn, NH, 2010).

______, 2010c, *SuperCivilizations: Civilizations as Superorganisms* (McMann-Fisher Publishing, Auburn, NH, 2010).

______, 2011a, *21st Century Natural Philosophy Of Ultimate Physical Reality* (McMann-Fisher Publishing, Auburn, NH, 2011).

______, 2011b, *All the Universe! Faster Than Light Tachyon Quark Starships & Particle Accelerators with the LHC as a Prototype Starship Drive Scientific Edition* (Pingree-Hill Publishing, Auburn, NH, 2011).

______, 2011c, *From Asynchronous Logic to The Standard Model to Superflight to the Stars* (Blaha Research, Auburn, NH, 2011).

______, 2012a, *From Asynchronous Logic to The Standard Model to Superflight to the Stars volume 2: Superluminal CP and CPT, U(4) Complex General Relativity and The Standard Model, Complex Vierbein General Relativity, Kinetic Theory, Thermodynamics* (Blaha Research, Auburn, NH, 2012).

______, 2012b, *Standard Model Symmetries, And Four And Sixteen Dimension Complex Relativity; The Origin Of Higgs Mass Terms* (Blaha Reasearch, Auburn, NH, 2012).

______, 2013a, *Multi-Stage Space Guns, Micro-Pulse Nuclear Rockets, and Faster-Than-Light Quark-Gluon Ion Drive Starships* (Blaha Research, Auburn, NH, 2013).

______, 2013b, *The Bridge to Dark Matter; A New Sister Universe; Dark Energy; Inflatons; Quantum Big Bang; Superluminal Physics; An Extended Standard Model Based on Geometry* (Blaha Reasearch, Auburn, NH, 2013).

______, 2014a, *Universes and Megaverses: From a New Standard Model to a Physical Megaverse; The Big Bang; Our Sister Universe's Wormhole; Origin of the Cosmological Constant, Spatial Asymmetry of the Universe, and its Web of Galaxies; A Baryonic Field between Universes and Particles; Megaverse Extended Wheeler-DeWitt Equation* (Blaha Reasearch, Auburn, NH, 2014).

______, 2014b, *All the Megaverse! Starships Exploring the Endless Universes of the Cosmos Using the Baryonic Force* (Blaha Research, Auburn, NH, 2014).

______, 2014c, *All the Megaverse! II Between Megaverse Universes: Quantum Entanglement Explained by the Megaverse Coherent Baryonic Radiation Devices – PHASERs Neutron Star Megaverse Slingshot Dynamics Spiritual and UFO Events, and the Megaverse Microscopic Entry into the Megaverse* (Blaha Research, Auburn, NH, 2014).

______, 2015a, *PHYSICS IS LOGIC PAINTED ON THE VOID: Origin of Bare Masses and The Standard Model in Logic, U(4) Origin of the Generations, Normal and Dark Baryonic Forces, Dark Matter, Dark Energy, The Big Bang, Complex General Relativity, A Megaverse of Universe Particles* (Blaha Research, Auburn, NH, 2015).

______, 2015b, *PHYSICS IS LOGIC Part II: The Theory of Everything, The Megaverse Theory of Everything, U(4)⊗U(4) Grand Unified Theory (GUT), Inertial Mass = Gravitational Mass, Unified Extended Standard Model and a New Complex General Relativity with Higgs Particles, Generation Group Higgs Particles* (Blaha Research, Auburn, NH, 2015).

______, 2015c, *The Origin of Higgs ("God") Particles and the Higgs Mechanism: Physics is Logic III, Beyond Higgs – A Revamped Theory With a Local Arrow of Time, The Theory of Everything Enhanced, Why Inertial Frames are Special, Universes of the Mind* (Blaha Research, Auburn, NH, 2015).

______, 2015d, *The Origin of the Eight Coupling Constants of The Theory of Everything: U(8) Grand Unified Theory of Everything (GUTE), S^8 Coupling Constant Symmetry, Space-Time Dependent Coupling Constants, Big Bang Vacuum Coupling Constants, Physics is Logic IV* (Blaha Research, Auburn, NH, 2015).

______, 2016a, *New Types of Dark Matter, Big Bang Equipartition, and A New U(4) Symmetry in the Theory of Everything: Equipartition Principle for Fermions, Matter is 83.33% Dark, Penetrating the Veil of the Big Bang, Explicit QFT Quark Confinement and Charmonium, Physics is Logic V* (Blaha Research, Auburn, NH, 2016).

______, 2016b, *The Periodic Table of the 192 Quarks and Leptons in The Theory of Everything: The U(4) Layer Group, Physics is Logic VI* (Blaha Research, Auburn, NH, 2016).

______, 2016c, *New Boson Quantum Field Theory, Dark Matter Dynamics, Dark Matter Fermion Layer Mixing, Genesis of Higgs Particles, New Layer Higgs Masses, Higgs Coupling Constants, Non-Abelian Higgs Gauge Fields, Physics is Logic VII* (Blaha Research, Auburn, NH, 2016).

______, 2016d, *Unification of the Strong Interactions and Gravitation: Quark Confinement Linked to Modified Short-Distance Gravity; Physics is Logic VIII* (Blaha Research, Auburn, NH, 2016).

______, 2016e, *MoND: Unification of the Strong Interactions and Gravitation II, Quark Confinement Linked to Large-Scale Gravity, Physics is Logic IX* (Blaha Research, Auburn, NH, 2016).

______, 2016f, *CQ Mechanics: A Unification of Quantum & Classical Mechanics, Quantum/Semi-Classical Entanglement, Quantum/Classical Path Integrals, Quantum/Classical Chaos* (Blaha Research, Auburn, NH, 2016).

______, 2016g, *GEMS: Unified Gravity, ElectroMagnetic and Strong Interactions: Manifest Quark Confinement, A Solution for the Proton Spin Puzzle, Modified Gravity on the Galactic Scale* (Pingree Hill Publishing, Auburn, NH, 2016).

______, 2016h, *Unification of the Seven Boson Interactions based on the Riemann-Christoffel Curvature Tensor* (Pingree Hill Publishing, Auburn, NH, 2016).

______, 2017a, *Unification of the Eleven Boson Interactions based on 'Rotations of Interactions'* (Pingree Hill Publishing, Auburn, NH, 2017).

______, 2017b, *The Origin of Fermions and Bosons, and Their Unification* (Pingree Hill Publishing, Auburn, NH, 2017).

______, 2017c, *Megaverse: The Universe of Universes* (Pingree Hill Publishing, Auburn, NH, 2017).

______, 2017d, *SuperSymmetry and the Unified SuperStandard Model* (Pingree Hill Publishing, Auburn, NH, 2017).

______, 2017e, *From Qubits to the Unified SuperStandard Model with Embedded SuperStrings: A Derivation* (Pingree Hill Publishing, Auburn, NH, 2017).

______, 2017f, *The Unified SuperStandard Model in Our Universe and the Megaverse: Quarks, ...* , (Pingree Hill Publishing, Auburn, NH, 2017).

______, 2018a, *The Unified SuperStandard Model and the Megaverse SECOND EDITION A Deeper Theory based on a New Particle Functional Space that Explicates Quantum Entanglement Spookiness (Volume 1)* (Pingree Hill Publishing, Auburn, NH, 2018).

______, 2018b, *Cosmos Creation: The Unified SuperStandard Model, Volume 2, SECOND EDITION* (Pingree Hill Publishing, Auburn, NH, 2018).

______, 2018c, *God Theory (*Pingree Hill Publishing, Auburn, NH, 2018).

______, 2018d, *Immortal Eye: God Theory: Second Edition* (Pingree Hill Publishing, Auburn, NH, 2018).

______, 2018e, *Unification of God Theory and Unified SuperStandard Model THIRD EDITION* (Pingree Hill Publishing, Auburn, NH, 2018).

______, 2019a, *Calculation of: QED α = 1/137, and Other Coupling Constants of the Unified SuperStandard Theory* (Pingree Hill Publishing, Auburn, NH, 2019).

______, 2019b, *Coupling Constants of the Unified SuperStandard Theory SECOND EDITION* (Pingree Hill Publishing, Auburn, NH, 2019).

______, 2019c, *New Hybrid Quantum Big_Bang–Megaverse_Driven Universe with a Finite Big Bang and an Increasing Hubble Constant* (Pingree Hill Publishing, Auburn, NH, 2019).

______, 2019d, *The Universe, The Electron and The Vacuum* (Pingree Hill Publishing, Auburn, NH, 2019).

______, 2019e, *Quantum Big Bang – Quantum Vacuum Universes (Particles)* (Pingree Hill Publishing, Auburn, NH, 2019).

______, 2019f, *The Exact QED Calculation of the Fine Structure Constant Implies ALL 4D Universes have the Same Physics/Life Prospects* (Pingree Hill Publishing, Auburn, NH, 2019).

______, 2019g, *Unified SuperStandard Theory and the SuperUniverse Model: The Foundation of Science* (Pingree Hill Publishing, Auburn, NH, 2019).

______, 2020a, *Quaternion Unified SuperStandard Theory (The QUeST) and Megaverse Octonion SuperStandard Theory (MOST)* (Pingree Hill Publishing, Auburn, NH, 2020).

______, 2020b, *United Universes Quaternion Universe - Octonion Megaverse* (Pingree Hill Publishing, Auburn, NH, 2020).

______, 2020c, *Unified SuperStandard Theories for Quaternion Universes & The Octonion Megaverse* (Pingree Hill Publishing, Auburn, NH, 2020).

______, 2020d, *The Essence of Eternity: Quaternion & Octonion SuperStandard Theories* (Pingree Hill Publishing, Auburn, NH, 2020).

______, 2020e, *The Essence of Eternity II* (Pingree Hill Publishing, Auburn, NH, 2020).

______, 2020f, *A Very Conscious Universe* (Pingree Hill Publishing, Auburn, NH, 2020).

______, 2020g, *Hypercomplex Universe* (Pingree Hill Publishing, Auburn, NH, 2020).

______, 2020h, *Beneath the Quaternion Universe* (Pingree Hill Publishing, Auburn, NH, 2020).

______, 2020i, *Why is the Universe Real? From Quaternion & Octonion to Real Coordinates* (Pingree Hill Publishing, Auburn, NH, 2020).

______, 2020j, *The Origin of Universes: of Quaternion Unified SuperStandard Theory (QUeST); and of the Octonion Megaverse (UTMOST)* (Pingree Hill Publishing, Auburn, NH, 2020).

______, 2020k, *The Seven Spaces of Creation: Octonion Cosmology* (Pingree Hill Publishing, Auburn, NH, 2020).

______, 2020l, *From Octonion Cosmology to the Unified SuperStandard Theory of Particles* (Pingree Hill Publishing, Auburn, NH, 2020).

Eddington, A. S., 1952, *The Mathematical Theory of Relativity* (Cambridge University Press, Cambridge, U.K., 1952).

Fant, Karl M., 2005, *Logically Determined Design: Clockless System Design With NULL Convention Logic* (John Wiley and Sons, Hoboken, NJ, 2005).

Feinberg, G. and Shapiro, R., 1980, *Life Beyond Earth: The Intelligent Earthlings Guide to Life in the Universe* (William Morrow and Company, New York, 1980).

Gelfand, I. M., Fomin, S. V., Silverman, R. A. (tr), 2000, *Calculus of Variations* (Dover Publications, Mineola, NY, 2000).

Giaquinta, M., Modica, G., Souchek, J., 1998, *Cartesian Coordinates in the Calculus of Variations* Volumes I and II (Springer-Verlag, New York, 1998).

Giaquinta, M., Hildebrandt, S., 1996, *Calculus of Variations* Volumes I and II (Springer-Verlag, New York, 1996).

Gradshteyn, I. S. and Ryzhik, I. M., 1965, *Table of Integrals, Series, and Products* (Academic Press, New York, 1965).

Heitler, W., 1954, *The Quantum Theory of Radiation* (Claendon Press, Oxford, UK, 1954).

Huang, Kerson, 1992, *Quarks, Leptons & Gauge Fields 2nd Edition* (World Scientific Publishing Company, Singapore, 1992).

Jost, J., Li-Jost, X., 1998, *Calculus of Variations* (Cambridge University Press, New York, 1998).

Kaku, Michio, 1993, *Quantum Field Theory*, (Oxford University Press, New York, 1993).

Kirk, G. S. and Raven, J. E., 1962, *The Presocratic Philosophers* (Cambridge University Press, New York, 1962).

Landau, L. D. and Lifshitz, E. M., 1987, *Fluid Mechanics 2nd Edition*, (Pergamon Press, Elmsford, NY, 1987).

Misner, C. W., Thorne, K. S., and Wheeler, J. A., 1973, *Gravitation* (W. H. Freeman, New York, 1973).

Rescher, N., 1967, *The Philosophy of Leibniz* (Prentice-Hall, Englewood Cliffs, NJ, 1967).

Rieffel, Eleanor and Polak, Wolfgang, 2014, *Quantum Computing* (MIT Press, Cambridge, MA, 2014).

Riesz, Frigyes and Sz.-Nagy, Béla, 1990, *Functional Analysis* (Dover Publications, New York, 1990).

Sagan, H., 1993, *Introduction to the Calculus of Variations* (Dover Publications, Mineola, NY, 1993).

Sakurai, J. J., 1964, *Invariance Principles and Elementary Particles* (Princeton University Press, Princeton, NJ, 1964).

Streater, R. F. and Wightman, A. S., 2000, *PCT, Spin, Statistics, and All That* (Princeton University Press, Princeton, NJ 2000).

Weinberg, S., 1972, *Gravitation and Cosmology* (John Wiley and Sons, New York, 1972).

Weinberg, S., 1995, *The Quantum Theory of Fields Volume I* (Cambridge University Press, New York, 1995).

Weinberg, S., 2000, *The Quantum Theory of Fields Volume III Supersymmetry* (Cambridge University Press, New York, 2000).

Weyl, H., 1950, *Space, Time, Matter* (Dover, New York, 1950).

Weyl, H., (Tr. S. Pollard et al), 1987, *The Continuum* (Dover Publications, New York, 1987).

INDEX

10 dimension space., 75
231 combinations, 116
anti-universes, 210
Asynchronous Logic, 230, 243
Axioms, 5
bags, 105
Bailin, 147
baryonic force, 243
baryonic gauge field, 210
Black Hole, 46, 229
Bogoliubov transformations, 158
boost transformation, 213, 214
bosonic string, 147
BQUeST, 36, 41, 57, 75, 89, 139, 140, 184, 189, 190
bright patches, 197
Cascading Quantum Probability Tweaks, 118
channels, xv, 110, 111, 115, 116, 117
Charmonium, 139, 157, 242
cloaked propagator, 147
coherent state, 163, 170
Coherent states, 223
Cold Spot, 198
complex, 141, 222
Complex General Coordinate Transformations, 175
Complex General Relativity, 169, 173, 182, 230, 231
complex quaternion space, 34, 54, 59, 186
Complex Special Relativity, 159
Complexon Feynman Propagator, 150
Computation, 95
condensation, 141
Connection, 119, 120, 121, 122, 126, 127, 128
conservation law, 143
Cosmic Microwave Background, 197
Cosmological Constant, 231
Cosmology, 137
Cosmos, iv, 118, 119, 120, 121, 122, 123, 124, 126, 127, 128, 129, 233, 242
Coulomb gauge, 154, 159, 221
coupling constants, 139, 156
CQMechanics, 156
Creation, 119, 233
Creationism, 126
Dark Energy, 140, 230
Dark Matter, 230, 231
Dimensional Interaction, 141
dimensions, 141, 223
Dirac equation, 212
Dirac matrices, 212
divergences, 147, 241
divine "spark, 119, 122
Durham University group, 198
electron, 210
ElectroWeak, 229, 230
Emanations, 110, 115
energy-time uncertainty relation, 151
Entropy, 168
EPR, 133, 134
equal time commutation relations, 144, 153, 221
Evolution, 126
Fermion group, 61, 63
fermion species, 212, 214
fermion triangle divergence, 155
Feynman propagators, 141, 146, 147, 166
fine structure constant, 242
Fock states, 222, 223
forms, 214
Fourier Wave Space, 130

Free Will, 118, 122, 123
functional, 129, 130, 199, 211
gates, 116, 117
gauge invariance, 143, 147
Gaussian, 141, 155, 225, 226
General Relativistic Reality group, 174, 176
General Relativistic Reality Group, 176
Generation group, 37, 62, 160, 169, 182, 188
Generation Group, 61, 62, 64, 181, 231
ghost, 141
Grand Unified Theory of Everything, 231
gravitational field fluctuations, 151
gravitational mass, 169, 170, 182
graviton, 152
Great Attractor, 197
Gupta-Bleuler gauge, 144
Hebrew alphabet, 115, 116
Higgs bosons, 107
Higgs Mechanism, 231, 241
Higgs microscope, 160
Higgs particles, 139, 156, 160, 161, 163, 165, 167, 168, 169, 170, 171, 182, 241
hypercomplex, 3, 7, 8, 9, 13
imaginary coordinates, 141, 147, 222, 223, 224
indices, 46, 80
inertial and gravitational mass, 179
inertial mass, 169, 170, 182
inertial reference frames, 165, 166
instantaneous action-at-a-distance, 133
instantaneous communication, 129
interactions, 241
Interdimensional Interaction, 141
internal dimensions, 46, 80
internal symmetries, 57
internal symmetry functionals, 130
ISIS, 243
Jacobian, 142, 149
Killing vector, 139, 156
Klein-Gordon field, 145
lagrangian, 143
Laniakea Galaxy Supercluster, 197
Layer group, 37, 160, 169, 179, 181, 182, 188
Layer groups, 61, 62
left-handed, 214, 215, 217, 218, 219, 220
left-handed neutrinos, 214
Leibniz, 119, 120, 235
LHC, iv, 230
Linguistics, 115
Lorentzian transformation, 174
mass scale, 141, 224
Mathematical Picture Language, 4, 9, 53
Maxiverse, v, vii, xiv, 1, 2, 20, 21, 22, 23, 28, 35, 72, 78, 89, 91, 93, 94, 95, 96, 103, 112, 113, 138
Megaverse, iv, 118, 127, 233, 234
minispace, 24, 28, 99
Minispace., 20, 99
MOST, iv, 7, 8, 47, 73, 204, 207, 233
Multiverse, 41
Negative Energy Scalar Particle States, 164
negative energy states, 170
octet, xi, 18, 19, 20
Octonion, iv, 233
Octonion Cosmology, iv, 22, 28, 91, 130, 137, 234
Octonion Octonion Octonion, 19, 20, 21, 22, 26, 95, 113, 114, 138
Octonion Octonion space, 91
octonions, 15
panentheistic, 119
paradoxes, 241
Parity Violation, 230
Pauli Exclusion Principle, 161
perturbation theory, 141
perturbation theory divergences, 42
photon, 222, 225
Planck mass, 141, 224

planckton field, 153
point space, 23, 24, 129, 130
Predestination, 122
principle value propagators, 166, 167
Pseudoquantum Field Theory, 139, 156
PseudoQuantum field Theory, 139
PseudoQuantum Field Theory, 156
PseudoQuantum formulation, 42
psiphi, 4, 10, 11
Pythagoras, 4, 5, 10
Pythagorean perfect number, 5
QED, 151, 224, 225
Quantum, 118, 122, 123, 124, 125, 126, 127, 233, 241, 243
quantum computers, 241
Quantum Dimensions, 141, 147, 222, 223, 224, 229
Quantum Entanglement, 135, 231
quantum fluctuations, 141
quantum functional, 133, 134
Quantum Gravity, 141, 151, 152, 229
Quantum Observer, 124
quark, 241
Quaternion Minispace, 99
quaternions, 13
QUeST, iv, 34, 54, 59, 186, 233, 244
radiation gauge, 144
Ranga-Ram Chary, 197
real dimensions, 33, 68, 184, 200
Reality group, 169, 175, 176, 182, 210
renormalization, 151
retarded propagators, 166, 167
right-handed, 216, 217, 218, 219, 220
Robertson-Walker metric, 241
S matrix, 8, 10
scalar field, 145, 150, 222
scalar universe particle, 42
Schrödinger's Cat, 124
seed fermion, 1, 41, 42, 44, 46, 47, 49, 50, 51, 56, 75, 76, 78, 80, 209
Sefer Yetzirah, i, 1, 2, 23, 27, 109, 110, 111, 114, 115, 116, 117, 137
Sefirot, 23
Shapley Attractor, 197
spaceless spaces, 23, 27, 130
Spaceless Spaces, 95
Space-Time Dependent Particle Masses, 168
Special Relativity, 229
species, 47, 61, 62, 73, 204, 207
Species Group Higgs Mechanism, 181
Species symmetry group, 42
spin, 241
spin states, 209
Spin-Momentum Functionals Space, 130
Spinoza, 119
Standard Model, 147, 151, 229, 241
stream, 46, 80
SU(3), 230
SuperStandard Model, iv, 118, 232, 233, 234
Superstring, 147, 229
SuperString theory, 89
Supersymmetric, 147
SuperSymmetry, 232
Superverse, 19, 22, 23, 25, 27, 101, 111, 112, 113, 115
symmetry breaking, 57
T Invariance, 166
T. Shanks et al, 198
tachyon spinor, 215
tachyon universe, 219
Tape of Fate, 122, 125, 126
Ten Space Cosmology, 130
Ten Spaces, 130
tetrackys, 4
Theory of Everything, 160, 170, 229, 231
Thermodynamics, 230
tiles, 57, 60, 61, 63
tiling, 57
time intervals, 151
translational invariance, 143
triangle divergence, 42, 76
triangle divergence., 42

two field formalism, 127
two-tier, 147, 148, 150, 151, 152, 226
Two-Tier, 155
Two-Tier coordinates, 42, 140
two-tier formalism, 142
Two-Tier propagators, 147
Two-Tier PseudoQuantum Field Theory, 158
Two-Tier Quantum Field Theory, 140
U(4), 230, 231
U(4) Species group, 43, 45, 58, 77, 79, 179
U(8), 231
Unified SuperStandard Model, iv, 118, 119, 120, 122, 127, 128, 233, 234
unitary, 214
universe particle, 155, 210, 211, 212, 213, 217, 219, 220, 225, 226, 227, 228
Unmoved Mover, 118, 119, 120, 121, 125, 126, 127
urfermion, 1, 23, 28, 75, 76, 78, 81, 83, 84, 86, 89, 91, 209
UST, 1, 2, 17, 22, 24, 36, 37, 42, 43, 51, 53, 56, 57, 58, 60, 61, 63, 64, 77, 95, 130, 137, 184, 188, 244
UTMOST, 68, 69, 71, 73, 74, 91, 200, 201, 204, 205, 206, 207, 208
vacuum fluctuations, 150, 151, 152
vierbein, 43, 46, 77, 78, 80, 173, 177
Web of Galaxies, 231
Weyl, H., 235
Wheeler-DeWitt equation, 209, 211, 243

About the Author

Stephen Blaha is a well-known Physicist and Man of Letters with interests in Science, Society and civilization, the Arts, and Technology. He had an Alfred P. Sloan Foundation scholarship in college. He received his Ph.D. in Physics from Rockefeller University. He has served on the faculties of several major universities. He was also a Member of the Technical Staff at Bell Laboratories, a manager at the Boston Globe Newspaper, a Director at Wang Laboratories, and President of Blaha Software Inc. and of Janus Associates Inc. (NH).

Among other achievements he was a co-discoverer of the "r potential" for heavy quark binding developing the first (and still the only demonstrable) non-Aeolian gauge theory with an "r" potential; first suggested the existence of topological structures in superfluid He-3; first proposed Yang-Mills theories would appear in condensed matter phenomena with non-scalar order parameters; first developed a grammar-based formalism for quantum computers and applied it to elementary particle theories; first developed a new form of quantum field theory without divergences (thus solving a major 60 year old problem that enabled a unified theory of the Standard Model and Quantum Gravity without divergences to be developed); first developed a formulation of complex General Relativity based on analytic continuation from real space-time; first developed a generalized non-homogeneous Robertson-Walker metric that enabled a quantum theory of the Big Bang to be developed without singularities at $t = 0$; first generalized Cauchy's theorem and Gauss' theorem to complex, curved multi-dimensional spaces; received Honorable Mention in the Gravity Research Foundation Essay Competition in 1978; first developed a physically acceptable theory of faster-than-light particles; first derived a composition of extremums method in the Calculus of Variations; first quantitatively suggested that inflationary periods in the history of the universe were not needed; first proved Gödel's Theorem implies Nature must be quantum; provided a new alternative to the Higgs Mechanism, and Higgs particles, to generate masses; first showed how to resolve logical paradoxes including Gödel's Undecidability Theorem by developing Operator Logic and Quantum Operator Logic; first developed a quantitative harmonic oscillator-like model of the life cycle, and interactions, of civilizations; first showed how equations describing superorganisms also apply to civilizations. A recent book shows his theory applies successfully to the past 14 years of history and to *new* archaeological data on Andean and Mayan civilizations as well as Early Anatolian and Egyptian civilizations.

He first developed an axiomatic derivation of the form of The Standard Model from geometry – space-time properties – The Unified SuperStandard Model. It unifies all the known forces of Nature. It also has a Dark Matter sector that includes a Dark ElectroWeak sector with Dark doublets and Dark gauge interactions. It uses quantum coordinates to remove infinities that crop up in most

interacting quantum field theories and additionally to remove the infinities that appear in the Big Bang and generate inflationary growth of the universe. It shows gravity has a MOND-like form without sacrificing Newton's Laws. It relates the interactions of the MOND-like sector of gravity with the r-potential of Quark Confinement. The axioms of the theory lead to the question of their origin. We suggest in the preceding edition of this book it can be attributed to an entity with God-like properties. We explore these properties in "God Theory" and show they predict that the Cosmos exists forever although individual universes (or incarnations of our universe) "come and go." Several other important results emerge from God Theory such a functionally triune God. The Unified SuperStandard Theory has many other important parts described in the Current Edition of *The Unified SuperStandard Theory* and expanded in subsequent volumes.

Blaha has had a major impact on a succession of elementary particle theories: his Ph.D. thesis (1970), and papers, showed that quantum field theory calculations to all orders in ladder approximations could not give scaling deep inelastic electron-nucleon scattering. He later showed the eigenvalue equation for the fine structure constant α in Johnson-Baker-Willey QED had a zero at $\alpha = 1$ not 1/137 by solving the Schwinger-Dyson equations to all orders in an approximation that agreed with exact results to 4th order in α thus ending interest in this theory. In 1979 at Prof. Ken Johnson's (MIT) suggestion he calculated the proton-neutron mass difference in the MIT bag model and found the result had the wrong sign reducing interest in the bag model. These results all appear in Physical Review papers. In the 2000's he repeatedly pointed out the shortcomings of SuperString theory and showed that The Standard Model's form could be derived from space-time geometry by an extension of Lorentz transformations to faster than light transformations. This deeper space-time basis greatly increases the possibility that it is part of THE fundamental theory. Recently, Blaha showed that the Weak interactions differed significantly from the Strong, electromagnetic and gravitation interactions in important respects while these interactions had similar features, and suggested that ElectroWeak theory, which is essentially a glued union of the Weak interactions and Electromagnetism, possibly modulo unknown Higgs particle features, be replaced by a unified theory of the other interactions combined with a stand-alone Weak interaction theory. Blaha also showed that, if Charmonium calculations are taken seriously, the Strong interaction coupling constant is only a factor of five larger than the electromagnetic coupling constant, and thus Strong interaction perturbation theory would make sense and yield physically meaningful results.

In graduate school (1965-71) he wrote substantial papers in elementary particles and group theory: The Inelastic E- P Structure Functions in a Gluon Model. Phys. Lett. B40:501-502,1972; Deep-Inelastic E-P Structure Functions In A Ladder Model With Spin 1/2 Nucleons, Phys.Rev. D3:510-523,1971; Continuum Contributions To The Pion Radius, Phys. Rev. 178:2167-2169,1969; Character Analysis of U(N) and SU(N), J. Math. Phys. 10, 2156 (1969); and The Calculation of the Irreducible Characters of the Symmetric Group in Terms of the

Compound Characters, (Published as Blaha's Lemma in D. E. Knuth's book: *The Art of Computer Programming Vols. 1 – 4*).

In the early 1980's Blaha was also a pioneer in the development of UNIX for financial, scientific and Internet applications: benchmarked UNIX versions showing that block size was critical for UNIX performance, developing financial modeling software, starting database benchmarking comparison studies, developing Internet-like UNIX networking (1982) and developing a hybrid shell programming technique (1982) that was a precursor to the PERL programming language. He was also the manager of the AT&T ten-year future products development database. His work helped lead to commercial UNIX on computers such as Sun Micros, IBM AIX minis, and Apple computers.

In the 1980's he pioneered the development of PC Desktop Publishing on laser printers and was nominated for three "Awards for Technical Excellence" in 1987 by PC Magazine for PC software products that he designed and developed.

Recently he has developed a theory of Megaverses – actual universes of which our universe is one – with quantum particle-like properties based on the Wheeler-DeWitt equation of Quantum Gravity. He has developed a theory of a baryonic force, which had been conjectured many years ago, and estimated the strength of the force based on discrepancies in measurements of the gravitational constant G. This force, operative in D-dimensional space, can be used to escape from our universe in "uniships" which are the equivalent of the faster-than-light starships proposed in the author's earlier books. Thus travel to other universes, as well as to other stars is possible.

Blaha also considered the complexified Wheeler-DeWitt equation and showed that its limitation to real-valued coordinates and metrics generated a Cosmological Constant in the Einstein equations.

The author has also recently written a series of books on the serious problems of the United States and their solution as well as a book on the decline of Mankind that will follow from current social and genetic trends in Mankind.

In the past twenty years Dr. Blaha has written over 80 books on a wide range of topics. Some recent major works are: *From Asynchronous Logic to The Standard Model to Superflight to the Stars, All the Universe!, SuperCivilizations: Civilizations as Superorganisms, America's Future: an Islamic Surge, ISIS, al Qaeda, World Epidemics, Ukraine, Russia-China Pact, US Leadership Crisis, The Rises and Falls of Man – Destiny – 3000 AD: New Support for a Superorganism MACRO-THEORY of CIVILIZATIONS From CURRENT WORLD TRENDS and NEW Peruvian, Pre-Mayan, Mayan, Anatolian, and Early Egyptian Data, with a Projection to 3000 AD,* and *Mankind in Decline: Genetic Disasters, Human-Animal Hybrids, Overpopulation, Pollution, Global Warming, Food and Water Shortages, Desertification, Poverty, Rising Violence, Genocide, Epidemics, Wars, Leadership Failure.*

He has taught approximately 4,000 students in undergraduate, graduate, and postgraduate corporate education courses primarily in major universities, and large companies and government agencies.

Recently he developed a quantum theory, The Unified SuperStandard Theory (UST), which describes elementary particles in detail without the difficulties of conventional quantum field theory. He found that the internal symmetries of this theory could be exactly derived from an octonion theory called QUeST. He further found that another octonion theory (UTMOST) describes the Megaverse. It can hold QUeST universes such as our own universe. It has an internal symmetry structure which is a superset of the QUeST internal symmetries.

www.ingramcontent.com/pod-product-compliance
Lightning Source LLC
LaVergne TN
LVHW062122200926
848484LV00007B/1736
* 9 7 8 1 7 3 5 6 7 9 5 5 6 *